Vector Calculus

Other titles in this series

Linear Algebra
R B J T Allenby

Mathematical Modelling
J Berry and K Houston

Discrete Mathematics
A Chetwynd and P Diggle

Particle Mechanics
C Collinson and T Roper

Ordinary Differential Equations
W Cox

Vectors in 2 or 3 Dimensions
A E Hirst

Numbers, Sequences and Series
K E Hirst

Groups
C R Jordan and D A Jordan

Statistics
A Mayer and A M Sykes

Probability
J H McColl

Calculus and ODEs
D Pearson

Analysis
P E Kopp

In preparation

Introduction to Non-Linear Equations
J Berry

Modular Mathematics Series

Vector Calculus

W Cox

Electronic Engineering and Computer Science
Aston University

A member of the Hodder Headline Group
LONDON • SYDNEY • AUCKLAND

To Dr Richard Wynne, with gratitude for all he
has done for my family, and wishing him a long
and happy retirement.

First published in Great Britain in 1998 by
Arnold, a member of the Hodder Headline Group,
338 Euston Road, London NW1 3BH

http://www.arnoldpublishers.com

British Library Cataloguing in Publication Data
A catalogue record for this book is available from the British Library

Library of Congress Cataloging-in-Publication Data
A catalog record for this book is available from the Library of Congress

ISBN 0 340 67741 4

Publisher: Nicki Dennis
Production Editor: Julie Delf
Production Controller: Sarah Kett

Typeset in 10/12 Times by
AFS Image Setters Ltd, Glasgow
Printed and bound in Great Britain by
J W Arrowsmith Ltd, Bristol

Contents

Series Preface

This series is designed particularly, but not exclusively, for students reading degree programmes based on semester-long modules. Each text will cover the essential core of an area of mathematics and lay the foundation for further study in that area. Some texts may include more material than can be comfortably covered in a single module, the intention there being that the topics to be studied can be selected to meet the needs of the student. Historical contexts, real life situations and linkages with other areas of mathematics and more advanced topics are included. Traditional worked examples and exercises are augmented by more open-ended exercises and tutorial problems suitable for group work or self-study. Where appropriate, the use of computer packages is encouraged. The first-level texts assume only the A-level core curriculum.

Professor Chris D. Collinson
Dr Johnston Anderson
Mr Peter Holmes

Preface

This book covers the basic principles and methods of multi-variable calculus from partial differentiation to vector calculus, including the integral theorems of Green, Gauss and Stokes. The book would be appropriate for a second-semester module building on a solid foundation in calculus and vectors such as may be found in Pearson's *Calculus and ODEs* and Hirst's *Vectors in 2 or 3 Dimensions* in the same series.

The approach of the book encourages the reader to anticipate concepts and methods by attempting problems in the text which lead into new material. These engage the student in the development of the subject and aid independent learning. There are also exercises at the end of each section, and further exercises complete each chapter. Solutions are provided. Essays and discussion points are included in the further exercises, and are suitable for group work. I would also encourage the reader to write summaries of the chapters – attempting to condense them into just a couple of hundred words gives a wider appreciation and overview of their content.

The range of material covered in this book is wide and there is not the space to treat all topics in sufficient detail to satisfy all tastes. However, there are a large number of books referred to in the Bibliography that can be used to explore deeper. The chapters have been divided so that the theoretical aspects have been separated from the methodology in the first part of the book. The reader primarily interested in applying multi-variable calculus, such as an engineer, can omit the theoretical chapters without much loss, if required.

Despite the title of the book, the introduction of vectors is left until over half-way through the book. This is because a solid foundation in the calculus of more than one variable is an essential prerequisite for vector calculus, and indeed it is all that some readers may require. I have also not included sections on traditional applications of vector calculus, such as fluid dynamics and electromagnetism. These subjects require significant physical background, and are less widely studied these days. At least those who do meet them will be equipped with the appropriate mathematical tools after working through this book.

Bill Cox
November 1997

Acknowlegements

I am always pleased to be able to thank my colleague Barry Martin for his careful and detailed reading of my work, and his suggestions for improvement. Similarly, I am ever grateful for the exceptional patience and high technical ability shown by Val Tyas in turning my scribbles into a manuscript, and Lynn Burton for producing the diagrams. I would also like to thank Richard Leigh for his copy editing whilst the book was in production. All remaining errors are mine.

1 • Introduction: A View From the Hill

1.1 Steepness in any direction

This introductory chapter gives an overview of the subject matter of this book, starting from what we already know about elementary calculus of a single variable. And where better to start than on a hill?

In elementary calculus of a single variable we often start by looking at a two-dimensional 'hill' – the graph of a function $y = f(x)$ against x, as shown in Fig. 1.1. At any point we define the slope of the curve to be the slope of the tangent to the curve at that point. This is given by the derivative of the function $f(x)$ at that point, if it exists. Of course, mathematically, this is a crude way to define the derivative – to do a proper job, we would use limits. However, for our present purposes it is sufficient to rely on a graphical representation.

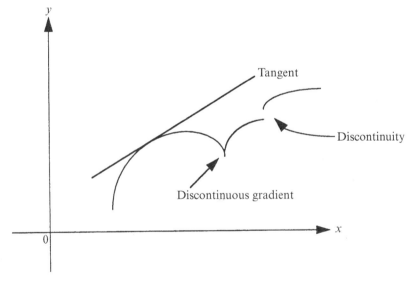

Fig. 1.1

Such a representation can be very informative. For example, if the curve has a break in it, or any sharp corners, then the tangent at such points is clearly not uniquely defined and so therefore neither is the derivative. Also, maximum and minimum points of the curve clearly correspond to points at which the tangent is horizontal – i.e. has zero gradient. Thus, at such points the derivative is zero. Again none of this is very rigorous, but it helps in forming our ideas. In the same way, similar diagrams can give us an insight into the differentiation of functions of more than one variable.

Now consider a function of two variables, $z = f(x,y)$. If we want to plot a 'graph' of this, then we need two axes $0x$ and $0y$, and for each pair of values (x,y) for which the function $f(x,y)$ is defined we plot the corresponding value of z in a third dimension. This generates a *surface in three dimensions* (*see* Fig. 1.2) as opposed to a curve in two dimensions for a function of a single variable – in other words a *real hill!*

PROBLEM I

Suppose you headed up the hill in the direction parallel to the xz-plane. How would you define the steepness of the hill at any point, and calculate the slope at that point? What if you headed in a direction parallel to the yz-plane?

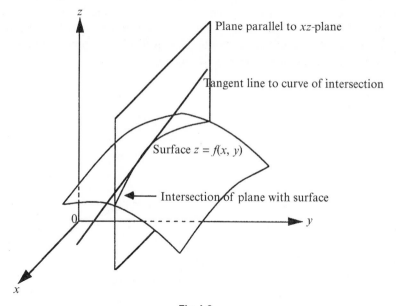

Fig. 1.2

A plane parallel to the xz-plane, through your position, will cut the surface in a curve, as shown in Fig. 1.2. As you walk up the hill, parallel to the xz-plane, you will walk along this curve, and the steepness at any point can be represented by the slope of the curve at that point. Since y does not vary at all along this curve, you can find the slope by 'differentiating' $f(x,y)$ with respect to x, treating y as a constant. This gives the *partial derivative of $z = f(x,y)$ with respect to x*, which is denoted by

$$\frac{\partial z}{\partial x}$$

to distinguish it from the 'ordinary derivative' dz/dx (∂ is called 'curly d'). Similarly, if you set off in a direction parallel to the yz-plane then you follow a curve on the surface cut by a plane through you and parallel to the yz-plane. The slope is now given by the partial derivative of z with respect to y, keeping x constant. This is denoted by $\partial z/\partial y$.

PROBLEM 2

> Now suppose you headed in some quite general direction. How would you determine the steepness on initially setting off?

You can see now that this is more difficult. You can fix a vertical plane (i.e. parallel to the z-axis) through your point of departure in the direction in which you set out, and you can see that this will cut the surface in a curve, the gradient of which, at your position, gives the initial steepness as you move off. But how to determine that gradient? I think you can appreciate – and we will confirm it in detail in Chapter 4 – that the slope will be some combination of $\partial z/\partial x$ and $\partial z/\partial y$. In fact, if you move a very small amount, equivalent to a displacement of Δx in the x-direction and Δy in the y-direction, then the resultant movement in the z-direction is approximately

$$\Delta z = \frac{\partial z}{\partial x}\Delta x + \frac{\partial z}{\partial y}\Delta y$$

So these first-order partial derivatives are clearly very important, and we will need to know their properties and how to calculate them. This is, fortunately, a fairly straightforward topic.

While we are at it, notice that a lot of what we have talked about so far is highly directional in nature, and so you should not be surprised at the use of vectors later in this book. In particular, the *tangent vector to a curve* is clearly going to be a very significant object.

EXERCISES ON 1.1

1. Evaluate $\partial f/\partial x$, $\partial f/\partial y$ for $f(x, y) = yx^2 \cos(x+y)$.
2. Starting at the point (1, 1, 2) on the 'hill' represented by the function $z = x^2 + y^2$, use the results of this section to calculate the approximate increase in vertical height in walking to the point with (x, y) coordinates (1.01, 1.01).

1.2 Reaching the top

So far, we have simply considered steepness at particular points on the hill. How do we know when we have reached the top, particularly if it is misty? Well, we hit a local *maximum* point – if we head off in any direction from such a point we will always go downwards.

PROBLEM 3

Characterize a maximum point geometrically and express this in terms of the values of the first derivatives.

Just as in functions of a single variable a maximum point is characterized by a horizontal tangent, so in the present case the maximum is characterized by a horizontal *tangent plane* – like the flat piece of a mortar-board. And, as the derivative is zero at such a point for functions of one variable, so both partial derivatives are zero in the present case. But this does not completely characterize a maximum value, of course. In functions of a single variable the second derivative must also be negative at such a point. Analogously, we will need *second-order partial derivatives* in order to discuss maxima (and other stationary values) in three dimensions.

Since $\partial z/\partial x$ is, just like z, a function of x and y, we can again partially differentiate it with respect to either. If we differentiate with respect to x then we obtain the *second partial derivative of z with respect to x*, denoted

$$\frac{\partial^2 z}{\partial x^2} = \frac{\partial}{\partial x}\left(\frac{\partial z}{\partial x}\right)$$

Similarly, we can differentiate with respect to y and obtain the *mixed derivative*

$$\frac{\partial^2 z}{\partial y \partial x} = \frac{\partial}{\partial y}\left(\frac{\partial z}{\partial x}\right)$$

Reversing this process and differentiating $\partial z/\partial y$ partially with respect to x and y respectively yields further second-order derivatives:

$$\frac{\partial^2 z}{\partial x \partial y} \text{ and } \frac{\partial^2 z}{\partial y^2}$$

As we will see in Chapter 4 , for appropriately well-behaved functions z, the mixed derivatives are equal:

$$\frac{\partial^2 z}{\partial y \partial x} = \frac{\partial^2 z}{\partial x \partial y}$$

From now on, in order to save paper, we will use a more succinct notation for the partial derivative of $f(x, y)$ with respect to x, which we write as f_x. Similarly, $\partial f/\partial y$ is written as f_y. Second derivatives are written as f_{xx}, f_{xy}, f_{yy}, and you can imagine the extensions to higher-order derivatives. Equality of mixed derivatives then becomes $f_{xy} = f_{yx}$.

EXERCISES ON 1.2

1. Consider the spherical surface

$$(x-1)^2 + (y-1)^2 + z^2 = 9$$

referred to a rectangular coordinate system $0xyz$. Give the points at which the horizontal tangent plane through the maximum value of z passes through the z-axis.

2. Given that

$$z = f(x, y) = \frac{1}{x+2y}$$

evaluate f_{xx}, f_{yy} and verify that $f_{xy} = f_{yx}$. What qualification must you make about the values of x and y?

1.3 Volumes in three dimensions

While wandering about on the hill we come across a quarry dug into the hillside. How could we determine the volume of material removed? What mathematical tools would we need for that? If you think back to functions of a single variable you will appreciate that we are effectively generalizing the calculation of the area under a curve, and for that we use integration.

PROBLEM 4

> Describe how the area under a curve between two points is found by integration. How may this be interpreted as division of the area into strips? How would you extend this to work out the volume under a surface over a given region? Express this in terms of integration.

The area under the curve $y = f(x)$ between the limits $x = a$ and $x = b > a$ is given by the integral

$$\int_a^b f(x)\, dx$$

and is illustrated in Fig. 1.3. This is equivalent to dividing the area into strips

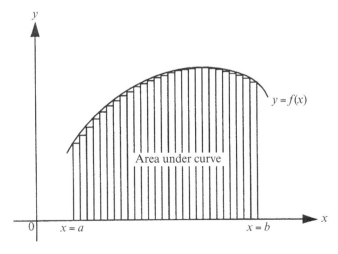

Fig. 1.3

parallel to the *y*-axis, as shown in the figure, totalling the area of the strips, and taking the strips to be thinner and thinner. If we now think of the *volume* under a surface, between *planes* $x=a$, $x=b$, $y=c$, and $y=d$, then you can imagine the volume divided into square columns parallel to the *z*-axis. We could first of all total up all the columns in a given row parallel to the *yz*-plane, say. This would be approximated by the thickness of the columns multiplied by the *area* under the curve cut by the plane of the columns with the surface. We know we can calculate such an area by *ordinary* integration with respect to *y* (*see* Fig. 1.4). If we now total up the volumes of all the 'slabs' parallel to the *yz*-plane, then this will require a second ordinary integration with respect to *x*. Putting all this together, we can see that finding such a volume is going to require the equivalent of *two ordinary integrations*. The following expression for the volume under the surface may now look reasonable:

$$V = \int_a^b \int_c^d f(x,y)\, \mathrm{d}y\, \mathrm{d}x$$

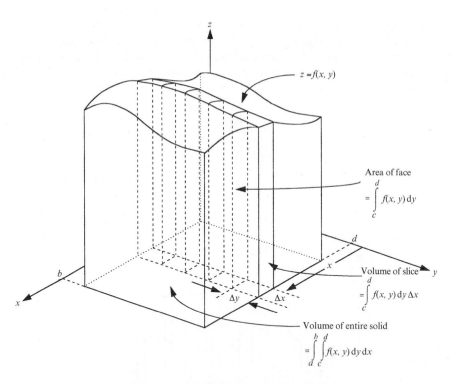

Fig. 1.4

This is an example of a *double integral* expressed in the form of a *repeated* or *iterated* integral over the rectangle $a \le x \le b$, $c \le y \le d$.

In this comparatively simple case of 'integration over a rectangle' we can evaluate the double integral by first integrating with respect to *y*, keeping *x* constant, and then integrating the result with respect to *x*.

1. As well as taking plates parallel to the *yz*-plane and evaluating their volumes first, we could also take plates parallel to the *xz*-plane. Repeat the discussion of the evaluation of the volume in this case. What do you deduce from your result?

2. Evaluate

$$\int_0^1 \int_2^3 (x + 2y)\, dy\, dx$$

and interpret the result.

1.4 Getting your bearings – vectors

So far, when discussing our hill climbing we have chosen to move in the directions of the axes of a fixed rectangular coordinate system. This is analogous to using, say, north–south directions on the compass, along with a vertical axis. But of course, most of our rambling will be in quite general directions – on bearings which can be represented by a vector at any given point. As soon as variations in two or three directions are being considered, vector notation comes into its own. Vectors enable us to represent mathematical objects and results in a way which is not dependent on any particular coordinate system. We will also need to be able to differentiate and integrate vectors and functions of vectors. Vectors have a rich structure, including two types of product – scalar and vector. We expect this to be reflected in the greater complexity of *vector calculus*. And indeed we get, for example, a number of different types of vector derivative operators – *grad*, *div* and *curl* (*see* Hirst, 1995). There are many interesting relations between these operators. These are studied in Chapters 8 and 9. We also get different types of integrals – *line*, *surface* and *volume integrals* – all designed to conform with the geometric structure of three (and indeed higher) dimensions. This is the subject of Chapters 7 and 10. Again, there are many interesting relations between these various types of integrals. In fact, there is one type of relation which reflects the deepest properties of theories about all physical phenomena, from the behaviour of the very smallest elementary particles to the cosmological workings of the universe. We can introduce this idea while wandering about on the hillside.

You will not go far in most mountain areas without coming across water and examples of fluid motion. In fact, mountains themselves are the result of collisions between tectonic plates floating on fluid magma – except that this is far more viscous than water and really a different type of fluid. Sometimes, if you are lucky, you may also see another, more spectacular, natural phenomenon in mountain areas – lightning, the violent discharge or flow of electricity between the atmosphere and earth.

All of these physical phenomena are the results of just two types of *force field* – the *gravitational* and the *electromagnetic*. And the underlying structure of these two fields is similar – the force is generated by some source (*mass* for gravitational, *electric charge* for electromagnetism), and acts to move some quantity producing *fluid flow*, or the *flow of electricity*. Even the terminology is the same. We are quite directly familiar with the importance of these two forces. Modern physics believes that in fact *all* physical phenomena can be explained in terms of just four funda-mental forces – the gravitational, electromagnetic, and two nuclear forces: the *weak* (radioactivity), and the *strong* (nuclear fusion). And all these forces are believed to have the same underlying theoretical structure: production by a 'charged' source, producing a force field which can induce motion – a *flow* or *current* – in a quantity on which it acts.

In fact, the underlying theory is so similar for these phenomena that they can be formulated in very similar mathematical languages. This has led to the search for a *unified field theory* in which all the four fundamental fields are incorporated into one unifying theory. Most physicists believe they have unified the electromagnetic and weak and strong nuclear forces. As yet they have not succeeded in unifying these with the most familiar force of all – gravity.

So how does this theoretical unification come about, and what has it got to do with the subject of this book? In fact, the mathematical thread through all of these field theories, which leads to the unification, is the extension of vector calculus to higher dimensions and more complicated spaces. And key in this is the idea of an *integral theorem*, which enables us to relate the *flow* or *flux* of a quantity out of a region to the amount of 'source' material within the region. In three dimensions the former is given by a double integral – a surface integral over the bounding surface of the region. The latter is given by a triple integral – a volume integral over the extent of the region. The relation between the two is what is called *Gauss's diver-gence theorem*, studied in Chapter 10.

Finally, back to the water. Some of the streams may have whirlpools. Toss a stick in and watch it circulate round the pool. Clearly, its speed of rotation has something to do with the 'angular momentum' of the rotating mass within the pool. This relationship can also be expressed by an integral theorem, called *Stokes's theorem*, again looked at in Chapter 10.

EXERCISES ON 1.4

1. The *gradient* of a function $f(x, y, z)$ is defined by

$$\text{grad } f = \frac{\partial f}{\partial x}\mathbf{i} + \frac{\partial f}{\partial y}\mathbf{j} + \frac{\partial f}{\partial z}\mathbf{k}$$

Evaluate:

(i) $\text{grad}(x^2y + y^2z)$ (ii) $\text{grad}(xyz)$

2. The *divergence* and *curl* of a vector function

$$\mathbf{F}(x, y, z) = F_1(x, y, z)\mathbf{i} + F_2(x, y, z)\mathbf{j} + F_3(x, y, z)\mathbf{k}$$

are defined respectively by

$$\text{div } \mathbf{F} = \frac{\partial F_1}{\partial x} + \frac{\partial F_2}{\partial y} + \frac{\partial F_3}{\partial z}$$

$$\text{curl } \mathbf{F} = \left(\frac{\partial F_3}{\partial y} - \frac{\partial F_2}{\partial z}\right)\mathbf{i} + \left(\frac{\partial F_1}{\partial z} - \frac{\partial F_3}{\partial x}\right)\mathbf{j} + \left(\frac{\partial F_2}{\partial x} - \frac{\partial F_1}{\partial y}\right)\mathbf{k}$$

Evaluate div and curl for the vector fields

(i) $xy\mathbf{i} + yz\mathbf{j} + zx\mathbf{k}$ (ii) $(x+y)\mathbf{i} + (y+z)\mathbf{j} + (x+z)\mathbf{k}$

FURTHER EXERCISES

1. Adapt the discussion of $\partial z/\partial x$ given in Section 1.1 to the case of $\partial z/\partial y$.

2. Evaluate $\partial z/\partial x$ and $\partial z/\partial y$ for the function $z = x^2 + y^2$. Hence determine the slope of the surface described by this function in the direction of the (i) x-axis, (ii) y-axis.

3. Find the approximate percentage change in $z = f(x, y) = 3x^2 y^2$ at the point $(1, 1)$ if we increase both x and y by 0.1%.

4. Evaluate all first- and second-order partial derivatives for the following functions:
    (i) $x^2 y^2$ (ii) $\cos(xy)$

5. Evaluate the first-order partial derivatives of the function
    $f(x, y) = 3xy - 6x - 3y + 7$
    at the point $(1, 2)$. What can you say about the surface $z = f(x, y)$ at this point?

6. Find the volume above the rectangle $1 \le x \le 2$, $0 \le y \le 2$ and under the surface $z = x^2 + y$. Evaluate the volume in two different ways and compare the results.

7. Evaluate the gradient of each of the functions
    (i) $3x + 2yz + xy$ (ii) $xy\cos(xz)$

8. Evaluate div and curl of the vector fields
    (i) $x^2 y^2 \mathbf{i} + 2xe^{yz}\mathbf{j} + z\mathbf{k}$ (ii) $e^{x+y}\mathbf{i} + e^{z+x}\mathbf{j} + e^{y+z}\mathbf{k}$

9. Essay or discussion topics:
    (i) The difference between the ordinary derivative dz/dx and the partial derivative $\partial z/\partial x$.
    (ii) The mathematical characteristics of a mountain pass between two peaks.
    (iii) Volume as a double integration.
    (iv) The vector operators grad, div and curl and their combinations.

2 • Functions of More Than One Variable

2.1 Functions of two variables

According to *Boyle's law*, the pressure, P, of a fixed mass of gas at constant temperature is inversely proportional to its volume, V, i.e. $P = C/V$, where C is a constant. In this case, the pressure is a *function of a single variable*, the volume, which we might write as $P = f(V)$. However, if the temperature, T, is also allowed to vary, then the *ideal gas law* tells us that the pressure, volume and temperature are related by $PV = nRT$, where nR is a constant. In this case, the pressure is a *function of two variables*, which we could write as $P = f(V,T)$. P varies as V, T or both vary, and it is clearly a more difficult task to keep track of the behaviour of P in the latter case, when both vary – even sketching a graph of the function $f(V,T)$ becomes problematical, for we would have to draw it in three dimensions. And imagine the trouble we are going to have with differentiation and integration! So, whereas in a book on the calculus of functions of a single variable the definition of a function might be dismissed in a couple of lines, here we are going to take more care over it. We will only consider real-valued functions in this book.

A brief reminder of the features of functions of a single variable will prepare us for the more complicated study of functions of two or more variables (*see* Pearson, 1996) A function, f, of a single real variable, x, is a rule for assigning a *unique* real number to *each* real number on which f acts – i.e. every element of its *domain*. The 'each' and the 'unique' are essential components of the definition of a function. They tell us that we must look carefully at the numbers or points on the real line involved in the definition of the function. Functions are defined on *sets* of such points. So, when we move on to functions of two variables, for example, we will be interested in points in a plane, defined by ordered pairs. Sets of such points are more complicated than those on a single real line, and so call for more careful treatment. Of course, in practice we usually deal with functions of a single variable in graphical terms, most usually referred to rectangular coordinates $0xy$ in a two-dimensional plane. Functions of two variables require such a plane simply to represent their domain, and there is a wider range of *coordinate systems* one can use in three dimensions. For example, plane polar coordinates (r, θ) may be usefully extended as either spherical polar coordinates or cylindrical polar coordinates in three dimensions. So, we have to give some attention to three- (and higher-)dimensional coordinate systems and the sketching of 'graphs' of functions of two variables. All this extra complexity for functions of more than one variable requires us to be rigorous about our definitions, terminology and notation, and it is the purpose of this chapter to set this out.

Suppose D is a non-empty set of ordered pairs (x,y). A *real-valued function f* on D is a mapping of each element of D onto the real numbers. D is called the *domain* of

f, and the set of values to which D is mapped is called the *range* of f. We write $z = f(x,y)$ where z is called the *image* of the ordered pair (x,y). In this case, x and y are called the *independent variables* of f, and z is the *dependent variable*.

The x and y in this definition are not necessarily coordinates in a rectangular coordinate system, but unless otherwise stated we will assume this for convenience. Then, D may be regarded as a set of points in the two-dimensional xy-plane, and z is plotted along the third axis perpendicular to this. Also, note that in the definition given above there is nothing special about the fact that we took ordered pairs. It is easily extended to n-tuples, $(x_1, x_2, \ldots, x_n)$, thereby defining a *function of n variables* in the obvious way.

Functions of several variables are named and classified in the same way as functions of a single variable. For example, a *polynomial*, $P(x, y)$, in two variables x and y is any finite combination of terms of the form $ax^m y^n$, where m and n are non-negative integers and a is a real number. A *rational function*, $R(x, y)$, in x and y is a ratio of two polynomials in x and y: $R(x, y) = P(x, y)/Q(x, y)$. The other standard elementary functions – circular, exponential, hyperbolic, logarithmic – can all appear in functions of several variables, of course.

PROBLEM I

> Consider the function $z = f(x, y) = x^2 + y^2$. Evaluate z at the points (x, y) for
>
> (i) $(0, 0)$ (ii) $(0, 1)$ (iii) $(1, 0)$ (iv) $(-2, 3)$ (v) $(-1, -2)$
>
> Sketch the function f in the domain
>
> $D = \{(x, y) : -2 \le x \le 2, -2 \le y \le 2\}$
>
> using rectangular coordinates $0xyz$.

The image of (i) $(0, 0)$ under the function f is $z = f(0, 0) = 0^2 + 0^2 = 0$. Similarly, (ii) $f(0, 1) = 0^2 + 1^2 = 1$, and you can check for yourself that (iii) is 1, (iv) is 13 and (v) is 5.

Figure 2.1 shows a sketch of $f(x, y)$. It has to be drawn in perspective, of course, and you may have trouble getting the shape right – do not worry if it takes you a few attempts. One way to tackle it is to consider the curves passing through points of equal height – i.e. equal z-value. In the present case, these curves are given by $x^2 + y^2 = c$, $z = c$, for different values of c. Clearly $c = 0$ corresponds to the origin; $c = 1$ corresponds to a circle of radius 1 centred on the z-axis in the $z = 1$ plane; and $c = 4$ corresponds to a circle of radius 2 centred on the z-axis in the $z = 4$ plane. We can also gain information about the shape of the surface by looking at the curves generated by the intersection of the surface with the xz- and yz-planes. In the xz-plane we have $y = 0$ and so the curve is the parabola $z = x^2$. Similarly, in the yz-plane we have $x = 0$ and the curve is the parabola $z = y^2$. Putting all this information together results in the sketch shown in Fig. 2.1.

The set of all points

$$(x, y, f(x, y))$$

in three-dimensional space, for all (x, y) in D, is called the *graph* of f, or the *surface* of $z = f(x, y)$. Thus, Fig. 2.1 represents the graph of $z = x^2 + y^2$.

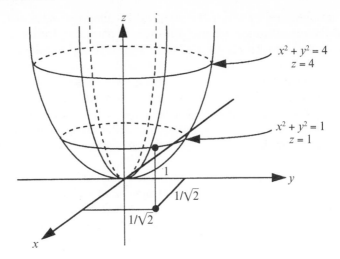

$x^2 + y^2 = 4$
$z = 4$

$x^2 + y^2 = 1$
$z = 1$

Fig. 2.1

If we project the curves of constant z-value considered above down onto the xy-plane, we obtain concentric circles about the origin, in the domain of the function f. These are called *level curves* or *contours* of f and form a two-dimensional 'plot' of f. They are obtained by plotting the sets of points for which $f(x, y)$ has a constant value, $f(x, y) = c$, say, in the xy-plane. This gives us the *level curve of height c*. As the name suggests, these curves help us in studying surfaces in the same way that contours on a map do.

The concept of a level curve for functions of two variables may be extended to that of a *level surface* for functions of three variables. Thus, if c is a constant then the equation $f(x, y, z) = c$ defines a surface in three dimensions called the *level surface of f with constant c*.

EXERCISES ON 2.1

1. Evaluate the following functions at the points indicated:
   (i) $f(x, y) = e^{-\sqrt{x^2+y^2}} \cos(x^2 + y^2)$
      (a) $(0, 0)$ (b) $(0, 1)$ (c) $(1, 0)$
   (ii) $f(x, y, z) = x^2 + y^2 + z^2$
      (a) $(-1, 1, 0)$ (b) $(1, 1, 1)$ (c) $(-1, 0, 2)$
2. Sketch the graph of the function $z = \sqrt{x^2 + y^2 + 1}$.
3. Sketch the level curves of height $-2, -1, 0, 1, 2$ for $z = x^2 - y^2$.

2.2 Sets of points in a plane

Having seen how difficult it is to view functions of more than one variable 'graphically', you can understand why it is that we will spend most of our time on functions of two variables only. In this case it is important to have a good appreciation of the domain of the function and this will consist of some set of points in the *xy*-plane. Thus we need to study such sets. In fact, the domains of functions are usually special types of sets.

In functions of a single variable we know that, while graphs of functions are useful pictorially, they cannot provide a rigorous basis for mathematical ideas. Thus, when we discuss such things as continuity and differentiability we use *intervals* and *limits*, on which the ideas of continuity and differentiability may be firmly based. In the case of a single variable, an interval is a set of numbers such that if two numbers are in the set then so is any number between them. When depicted on the real line, this corresponds to a section of the line which has no holes in it. Such intervals form the basic units in terms of which we discuss the concepts of analysis of a function of a single variable. They are expressed symbolically by the expression

$$|x - x_0| < \varepsilon$$

which represents the set of all points in the range $(x_0 - \varepsilon, x_0 + \varepsilon)$ – this is called an *open interval*. Similarly, $|x - x_0| \leq \varepsilon$ represents a *closed interval* about x_0, i.e. all points in the range $[x_0 - \varepsilon, x_0 + \varepsilon]$.

This symbolism suggests an obvious extension to points in a plane – and indeed to any space in which we have a concept of *distance*. In a plane, the generalization of the interval is a *disc*, and this motivates the following definitions.

The *open disc* of radius r and centred at (x_0, y_0) is the set of points, or ordered pairs, defined by

$$|(x, y) - (x_0, y_0)| = \sqrt{(x - x_0)^2 + (y - y_0)^2} < r$$

i.e. the set of points enclosed within the circle of radius r centred on (x_0, y_0). Similarly, the *closed disc* of radius r, centred on (x_0, y_0), is the set defined by

$$|(x, y) - (x_0, y_0)| = \sqrt{(x - x_0)^2 + (y - y_0)^2} \leq r$$

The *boundary* of such a disc is given by the circle

$$|(x, y) - (x_0, y_0)| = \sqrt{(x - x_0)^2 + (y - y_0)^2} = r$$

A *neighbourhood* of a point (x_0, y_0) is an open disc centred on the point (or more generally any open set containing such a disc).

PROBLEM 2

Write down the open and closed discs of radius 2, centred on (2, 3). Give the boundary of the disc. Give an example of a neighbourhood of the point (2, 3).

The open disc is defined by

$$\sqrt{(x - 2)^2 + (y - 3)^2} < 2$$

The closed disc is

$$\sqrt{(x - 2)^2 + (y - 3)^2} \leq 2$$

The boundary is

$$\sqrt{(x-2)^2+(y-3)^2}=2$$

A neighbourhood of $(2, 3)$ is any set of points defined by

$$\sqrt{(x-2)^2+(y-3)^2}<r$$

for some positive value of r.

As noted above, these ideas may be extended to three, or indeed more, dimensions by extending the distance measure or *metric* appropriately. The 'disc' becomes replaced by a *ball* – otherwise the terminology is unchanged.

Discs and balls are examples of bounded sets. A *bounded set* in a plane is a set which can be entirely enclosed within a rectangle of finite size – otherwise it is *unbounded*. Similarly for bounded sets in three, and indeed higher, dimensions.

Discs, balls and neighbourhoods can be used to describe the characteristics of quite general sets. Thus, if S is any set of points in the xy-plane, then we say a point (x_0, y_0) is an *interior point* of S if it can be surrounded by a disc of positive radius which contains only points of S. (x_0, y_0) is called a *boundary point* of S if every circular disc with positive radius, centred on (x_0, y_0), contains points in S and points not in S. Figure 2.2 will perhaps help make sense of these definitions, which extend to sets in three dimensions merely by replacing discs by balls.

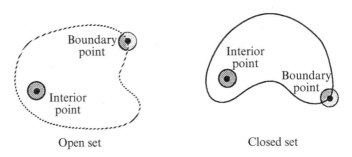

Open set   Closed set

Fig. 2.2

We can now generalize the definitions of open and closed discs to arbitrary sets in the xy-plane. Thus, a set S in 2-space or 3-space is called *open* if it contains none of its boundary points and *closed* if it contains all of its boundary points. The set of all points of the xy-plane is regarded as both open and closed – and similarly for all the points in 3-space.

The term 'region' is often used rather loosely to refer to sets of points in the plane. In fact, this is a very specific type of set, which corresponds precisely to intervals in analysis of functions of a single variable. Thus, an interval is a connected segment of the real line – there are no holes in it. In general, a set S in $\mathbb{R}^2$ or $\mathbb{R}^3$ is *connected* if any two points can be joined by a piecewise smooth curve enclosed entirely within S. A *region* in $\mathbb{R}^2$ or $\mathbb{R}^3$ is a connected set, which may be open or closed.

1. Classify the following sets in $\mathbb{R}^2$ and $\mathbb{R}^3$ as open, closed or neither. In each case give examples of an interior point and a boundary point.
   (i) $\{(x, y) : x^2 + y^2 \leq 2\}$
   (ii) $\{(x, y) : x^2 + y^2 > 1\}$
   (iii) $\{(x, y) : 1 < x^2 + y^2 \leq 3\}$
   (iv) $\{(x, y, z) : 0 < z < 2 - y \text{ and } 0 < x < 1\}$
   (v) $\{(x, y, z) : x^2 + y^2 + z^2 \leq 2\}$
   (vi) $\{(x, y) : x^2 + y^2 < 1\} \cup \{(x, y) : x^2 + y^2 > 2\}$
2. Classify the sets in Exercise 1 as bounded, unbounded, connected, or a region.

2.3 Three-dimensional coordinate systems

I will only give a brief reminder of the main types of coordinate systems used in three dimensions. Further details may be found in Hirst (1995). Note, however, that our notation differs somewhat from that used in Hirst. This section is just to introduce terminology which will be used later.

● *Rectangular coordinate system in three-dimensional space* ——

This consists of three mutually perpendicular axes intersecting each other in a single point, the *origin*. It is sometimes referred to as the *Cartesian coordinate system*. The scales on the coordinate axes are normally measured by variables labelled x, y, z. If you have seen such coordinate systems before, then you will almost certainly automatically label them as shown in Fig. 2.3(a). This is in fact what is referred to as a *right-handed* system. If you cup your right hand such that the fingers curve from the positive x-axis towards the positive y-axis, then the thumb points along the positive z-axis. Figure 2.3(b) illustrates a system which is not right-handed, but *left-handed*. We will only use right-handed coordinate systems.

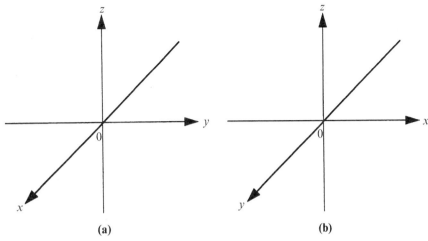

(a)   (b)

Fig. 2.3

The coordinate axes determine three *coordinate planes*: the *xy-plane*, *xz-plane* and *yz-plane*. Each point P in 3-space can be labelled uniquely by three numbers given by the intersection of planes through P, parallel to the coordinate planes with the coordinate axes. These points of intersection are called the *x*-coordinate, *y*-coordinate, and *z*-coordinate of P, and we denote the point by $P(x, y, z)$. The surfaces $x = x_0$, $y = y_0$, $z = z_0$ of which the coordinate planes are special cases (x_0, y_0, z_0 all zero) are called *constant surfaces*. They are planes parallel to the coordinate planes. The coordinate planes divide 3-space into eight equal parts called *octants*. The octant with all coordinates positive is called the *first octant*.

The distance between two points $P_1(x_1, y_1, z_1)$ and $P_2(x_2, y_2, z_2)$ referred to a rectangular coordinate system is given by applying Pythagoras's theorem twice:

$$d = \sqrt{(x_1 - x_2)^2 + (y_1 - y_2)^2 + (z_1 - z_2)^2}$$

The midpoint of the line segment joining these two points is given by

$$\left(\frac{1}{2}(x_1 + x_2), \frac{1}{2}(y_1 + y_2), \frac{1}{2}(z_1 + z_2) \right)$$

Notice how both of these results are 'obvious' generalizations of the corresponding results in a plane.

• Cylindrical coordinates in $\mathbb{R}^3$

The *cylindrical coordinates* (r, θ, z) of a point P in $\mathbb{R}^3$ are shown in Fig. 2.4(a). r and z are unbounded coordinates, but θ, being an angle, is limited to $0 \le \theta < 2\pi$.

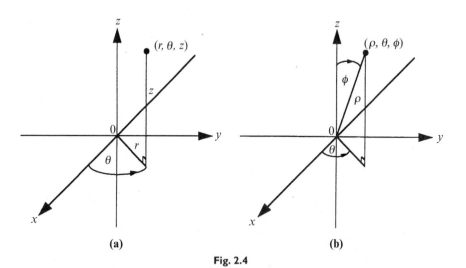

(a)   (b)

Fig. 2.4

The constant surfaces in this case are shown in Fig. 2.5(a). The surface $r = r_0$ is a cylinder of radius r_0 centred on the z-axis. The surface $\theta = \theta_0$ is a half-plane attached along the z-axis and making angle θ_0 with the positive x-axis. The surface $z = z_0$ is a horizontal plane through $z = z_0$ on the z-axis.

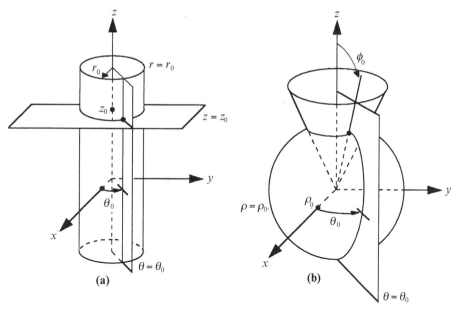

Fig 2.5

● *Spherical coordinates in* $\mathbb{R}^3$

The spherical coordinates (ρ, θ, ϕ) of a point P in $\mathbb{R}^3$ are shown in Fig. 2.4(b). ρ is an unbounded coordinate defining the radial distance from the origin, θ is angular, $0 \leq \theta < 2\pi$, and ϕ is also angular but only needs to range over $0 \leq \phi \leq \pi$ (why?). The surface $\rho = \rho_0$ is a sphere centred on the origin with radius ρ_0. The surface $\theta = \theta_0$ is a half-plane attached along the z-axis, making an angle of θ_0 with the positive x-axis. The surface $\phi = \phi_0$ consists of all points from which a line segment to the origin makes an angle ϕ_0 with the positive z-axis. Depending on whether $0 < \phi_0 < \pi/2$, or $\pi/2 < \phi_0 < \pi$, this will be a cone opening upwards or opening downwards. (If $\phi_0 = \pi/2$, then the cone is flat and the surface is the xy-plane.) These constant surfaces are shown in Fig. 2.5(b). Cylindrical/spherical coordinates are most useful in problems with cylindrical/spherical symmetry.

To convert between the different coordinate systems we use the transformations shown in Table 2.1, which we will need in Chapter 7.

Table 2.1

Conversion	Formulae
$(r, \theta, z) \rightarrow (x, y, z)$	$x = r \cos \theta,\ y = r \sin \theta,\ z = z$
$(x, y, z) \rightarrow (r, \theta, z)$	$r = \sqrt{x^2 + y^2},\ \tan \theta = y/x,\ z = z$
$(\rho, \theta, \phi) \rightarrow (r, \theta, z)$	$r = \rho \sin \phi,\ \theta = \theta,\ z = \rho \cos \phi$
$(r, \theta, z) \rightarrow (\rho, \theta, \phi)$	$\rho = \sqrt{r^2 + z^2},\ \theta = \theta,\ \tan \phi = r/z$
$(\rho, \theta, \phi) \rightarrow (x, y, z)$	$x = \rho \sin \phi \cos \theta,\ y = \rho \sin \phi \sin \theta,\ z = \rho \cos \phi$
$(x, y, z) \rightarrow (\rho, \theta, \phi)$	$\rho = \sqrt{x^2 + y^2 + z^2},\ \tan \theta = y/x,\ \cos \phi = z/\sqrt{x^2 + y^2 + z^2}$

EXERCISES ON 2.3

1. Prove the results for the distance between two points and the midpoint of a line.
2. Prove the transformation results given in Table 2.1.

2.4 Sketching graphs in three dimensions

It is much more difficult to sketch graphs in three dimensions than in two. We have few simple surfaces with which we are familiar, and are virtually restricted to the generalizations of the conic sections – the so-called *quadric surfaces*. There are, of course, many computer packages which generate the graphs of functions of two variables, but it is still useful to develop skills of sketching by hand. This can be done by considering the items in the following checklist:

- *Symmetry*: take advantage of any symmetry evident.
- *Boundedness*: consider whether the surface is bounded or unbounded.
- *Intercepts*: look for points where the surface intercepts the coordinate axes.
- *Centres*: look for a *centre* where appropriate (e.g. centre of a sphere).
- *Interesting points*: look for any other interesting points such as maxima or minima, or saddle points.
- *Traces*: look at intersections with the coordinate planes (the *traces*).
- *Sections*: look at intersections with other interesting planes.

You may not need to consider all these items, and not necessarily in the order given, but they at least provide a guide for sketching surfaces.

PROBLEM 3

> Sketch the surface
>
> $$\frac{x^2}{a^2} + \frac{y^2}{b^2} + \frac{z^2}{c^2} = 1$$
>
> (this is called an ellipsoid and is a particular example of a quadric surface – *see* Section 2.5).

Because the equation is invariant under any change of sign in x, y or z, the surface is symmetric about the xy-, xz- or yz-planes. The ellipsoid is centred at the origin. Since x, y, z, a, b, c are all real, each of the squared terms on the left-hand side is positive, and since they must always total to at most one, we must have $|x| \le |a|$, $|y| \le |b|$ and $|z| \le c$ – i.e. the surface is bounded by a cuboid with sides a, b, c. The intercepts on the axes are $x = \pm a$, $y = \pm b$, $z = \pm c$. These points are called *vertices*. There are no further points of special interest. All three traces ($x = 0$, $y = 0$, $z = 0$) are ellipses, as are all sections parallel to the coordinate planes. a, b, c are called the *semi-axes* of the ellipsoid. If any two are equal we have an *ellipsoid of revolution*. If all three are equal we have a *sphere*. Putting all this information together yields the sketch shown in Fig. 2.6.

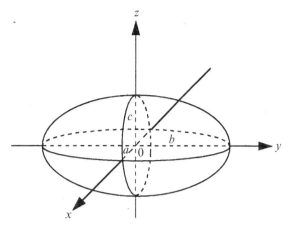

Fig. 2.6

Sketch the following surfaces:
(i) The hyperboloid of one sheet

$$\frac{x^2}{a^2} + \frac{y^2}{b^2} - \frac{z^2}{c^2} = 1$$

(ii) The hyperboloid of two sheets

$$\frac{x^2}{a^2} + \frac{y^2}{b^2} - \frac{z^2}{c^2} = -1$$

2.5 The quadric surfaces: project

The examples given in the previous section are particular cases of a very important general class of surfaces in three dimensions called the *quadric surfaces*. These are defined by equations of the form

$$Ax^2 + By^2 + Cz^2 + Dxy + Exz + Fyz + Hx + Iy + Jz + K = 0$$

where the coefficients A, B, ..., K are constants. You can get an idea of why such second-degree functions are important by thinking of the simple second-degree Taylor approximation of a function of a single variable

$$f(x) \approx A + Bx + Cx^2$$

Such second-order approximations are much used in applied mathematics. In such cases the behaviour of the function is dominated by the terms of second degree. The same applies to functions of more than one variable, and the expression occurring in the quadric is the most general second-degree expression in three variables, hence its importance.

It turns out that by suitable translations and rotations of the coordinate axes the quadrics can be classified into just nine distinct non-trivial types.

PROBLEM 4

Complete the sketches of the graphs of the following nine types of quadric surface given in standard form: (i), (ii), (iii) have already been done in Section 2.4, the remainder are left as exercises. The graphs may be found in most books on advanced calculus (e.g. Thomas and Finney, 1996).

(i) The ellipsoid

$$\frac{x^2}{a^2} + \frac{y^2}{b^2} + \frac{z^2}{c^2} = 1$$

(ii) The hyperboloid of one sheet

$$\frac{x^2}{a^2} + \frac{y^2}{b^2} - \frac{z^2}{c^2} = 1$$

(iii) The hyperboloid of two sheets

$$\frac{x^2}{a^2} + \frac{y^2}{b^2} - \frac{z^2}{c^2} = -1$$

(iv) The quadric cone

$$\frac{x^2}{a^2} + \frac{y^2}{b^2} = z^2$$

(v) The elliptic paraboloid

$$\frac{x^2}{a^2} + \frac{y^2}{b^2} = z$$

(vi) The hyperbolic paraboloid

$$\frac{x^2}{a^2} - \frac{y^2}{b^2} = z$$

(vii) The parabolic cylinder

$$y^2 = 4ax$$

(viii) The elliptic cylinder

$$\frac{x^2}{a^2} + \frac{y^2}{b^2} = 1$$

(ix) The hyperbolic cylinder

$$\frac{x^2}{a^2} - \frac{y^2}{b^2} = 1$$

FURTHER EXERCISES

1. Evaluate each of the following functions at the points indicated:

(i) $x \cos y$ $(0, 0), (1, 0), (\pi, \pi), (-1, \pi/2)$

(ii) $\frac{x^2 + y^2}{z^2}$ $(0, 0, 1), (-1, 2, 4), (1, 1, 2), (-1, -1, -1)$

(iii) $e^{x+y} \sin(yz)$ $(0, 0, 0), \left(1, \frac{\pi}{2}, 2\right), \left(\ln 2, \ln 2, \frac{\pi}{2\ln 2}\right), \left(1, -1, \frac{\pi}{4}\right)$

2. Find the maximum domain and range of the given functions:
   (i) $f(x, y) = \cos(x + y)$
   (ii) $f(x, y) = \sqrt{2 + x + y}$
   (iii) $f(x, y) = y/x$
   (iv) $f(x, y) = 2e^x + 3e^y$
   (v) $f(x, y) = \sqrt{x^2 + y^2}$
   (vi) $f(x, y) = x + y$
   (vii) $f(x, y) = \ln(1 - x^2 + y)$
   (viii) $f(x, y) = \sqrt{\dfrac{x + y}{x - y}}$
   (ix) $g(x, y, z) = \sqrt{x + 2y + 3z}$
   (x) $g(x, y, z) = z/(x + y)$
   (xi) $g(x, y, z) = \sin^{-1}[(x + z)/y]$
   (xii) $g(x, y, z) = \ln(x + y + 2x + 1)$

3. Sketch the surfaces generated by the following equations:
   (i) $z = x - y^2$
   (ii) $z = 3x^2 + 3y^2$
   (iii) $x^2 + 4y^2 - 16z^2 = 0$
   (iv) $9y^2 - 4x^2 - 36z^2 - 36 = 0$

4. Describe the level curves of the following functions:
   (i) $z = x - y;\ z = -4, -2, 0, 2, 4$
   (ii) $z = \dfrac{x}{x + y};\ z = -1, 0, 1, 2$
   (iii) $z = \sqrt{1 + x + y};\ z = 0, 1, 5, 10$
   (iv) $z = \sqrt{1 - x^2 - 4y^2};\ z = 0, \dfrac{1}{4}, \dfrac{1}{2}, 1$

5. Write equations defining the following sets of points:
   (i) The open disc centred at $(2, -1)$ with radius 2
   (ii) The closed disc centred at $(-1, -1)$ with radius 4
   (iii) The open ball centred at $(-1, 2, 0)$ with radius 3
   (iv) The closed ball centred at $(1, 1, 0)$ with radius 1

6. In the following, specify the interior and the boundary of the set. State whether the set is open, closed or neither.
   (i) $\{(x, y) : 1 \le x \le 3, 2 \le y \le 4\}$
   (ii) $\{(x, y) : a < x < b, c < x < d\}$
   (iii) $\{(x, y) : 1 < x^2 + y^2 < 9\}$
   (iv) $\{(x, y) : 2 < x^2 + y^2 \le 3\}$
   (v) $\{(x, y) : 1 < x^2 + 2y^2 < 5\}$
   (vi) $\{(x, y) : y \le x^2\}$
   (vii) $\{(x, y, z) : 1 < x < 2, 2 < y < 3, 0 < z < 1\}$
   (viii) $\{(x, y, z) : 1 < x \le 2, 2 \le y < 3, 0 \le z \le 1\}$
   (ix) $\{(x, y, z) : x^2 + y^2 \le 4, 0 \le z \le 3\}$
   (x) $\{(x, y, z) : (x - 1)^2 + (y + 2)^2 + (z - 3)^2 < 4\}$

7. Write the following equations in cylindrical coordinates:
   (i) $x^2 + y^2 + z^2 = 4$
   (ii) $x^2 + y^2 = 1$
   (iii) $z = 3\sqrt{x^2 + y^2}$
   (iv) $x^2 + z^2 = 9$

8. Complete the following table:

Rectangular coordinates	Cylindrical coordinates	Spherical coordinates
(i) $(1, 1, 1)$		
(ii)		$(4, \pi/3, \pi/4)$
(iii)	$(2, 2\pi/3, 6)$	
(iv) $(\sqrt{2}, -\sqrt{2}, 1)$		
(v)		$(2\sqrt{2}, \pi/2, 3\pi/2)$
(vi)	$(4, \pi/3, -3)$	

9. Find an equation in cylindrical coordinates of the surface whose equation in rectangular coordinates is $z = x^2 + y^2 + 2x + y$.

10. Find an equation in spherical coordinates of the surface whose equation in rectangular coordinates is $x^2 + y^2 + (z-1)^2 = 1$.

11. Express the following equations given in cylindrical and spherical coordinates respectively, in terms of rectangular coordinates:
    (i) $r = 4 \sin \theta$ (ii) $\rho \sin \phi = 2 \cos \theta$

12. Interpret each of the following equations in spherical coordinates geometrically:
    (i) $\rho \sin \phi = 1$ (ii) $\rho \cos \phi = 1$

13. Identify the following surfaces:
    (i) $x^2 + 9y^2 - 4z^2 = 0$ (ii) $x - 4y^2 = 0$
    (iii) $5x^2 + 2y^2 - 6z^2 + 10 = 0$ (iv) $x^2 + y^2 + z^2 - 4 = 0$
    (v) $x^2 + 2y^2 - z = 0$ (vi) $x - 4y^2 + 2z^2 = 0$

14. Identify and sketch the following surfaces:
    (i) $9y^2 + z^2 - 16 = 0$ (ii) $y^2 + z = 0$
    (iii) $y^2 - 4x^2 = 4$ (iv) $x^2 + 9y^2 - 25z = 0$
    (v) $x^2 + 9y^2 - 25z^2 = 0$ (vi) $x^2 - 9y^2 - 25z^2 = 25$
    (vii) $x^2 + 9y^2 - 25z = 0$ (viii) $x^2 + y^2 - 4z = 0$
    (ix) $x^2/4 + y^2/16 + z^2/9 = 1$ (x) $x^2 + y^2 - z^2/9 = 1$
    (xi) $x^2 + y^2/4 - z^2 = -1$ (xii) $z = x^2/9 + y^2/4$
    (xiii) $z = x^2/9 - y^2/4$

15. Essay or discussion topics:
    (i) The principal differences between a function of a single variable and functions of two or more variables.
    (ii) Intervals, discs, balls: what do they have in common as sets of points?
    (iii) Sketching surfaces in three-dimensional coordinate systems, including cylindrical and spherical.

3 • Limits and Continuity: Analytical Aspects

3.1 The case of a single variable

Particularly in limestone country, you will know to look out for potholes and crevices when wandering over hillsides. In the same way, 'hills' generated by functions can have discontinuities in them. In the case of functions of a single variable we approach such problems by means of limits. We can do the same in the case of functions of two or more variables, but first let us review the case of a function of a single variable

$$y = f(x)$$

Recall that the purpose of defining a limit of such a function is so that we can discuss the concepts of continuity and differentiability in a symbolic mathematical way.

PROBLEM I

Review the definitions of the single-sided limits

$$\lim_{x \to a^-} f(x) \quad \text{and} \quad \lim_{x \to a^+} f(x)$$

of $f(x)$ at a 'from below' and 'from above', respectively.

Explain how these limits can be used to discuss the continuity of $f(x)$ at $x = a$. Discuss the definition and existence of the derivative, $f'(a)$, of $f(x)$ at $x = a$, in terms of limits.

Suppose f is defined on some open interval of values less than a. Then the *limit of f as x tends to a from below* (or the *left-sided limit*) is written

$$L = \lim_{x \to a^-} f(x)$$

and exists if, given any number $\varepsilon > 0$, we can find a number $\delta > 0$ such that $f(x)$ satisfies

$$|f(x) - L| < \varepsilon$$

whenever x satisfies $a - \delta < x < a$.

Similarly, the *limit of f as x tends to a from above* (or the *right-sided limit*) is written

$$L = \lim_{x \to a^+} f(x)$$

and exists provided for any number $\varepsilon > 0$ we can find a $\delta > 0$ such that $|f(x) - L| < \varepsilon$ whenever x satisfies $a < x < a + \delta$.

These left- and right-sided limits are both necessary in the discussion of continuity of a function of a real variable, since they may not necessarily be the same. If they *are* the same, then we define their common value, L, as the *limit of f as x tends to a*, and write it

$$L = \lim_{x \to a} f(x) = \lim_{x \to a^-} f(x) = \lim_{x \to a^+} f(x)$$

We say that $f(x)$ is *continuous at* $x = a$ if both $f(a)$ and $\lim_{x \to a} f(x)$ exist and they are equal:

$$\lim_{x \to a} f(x) = f(a)$$

The derivative of $f(x)$ at $x = a$, which we denote by $f'(a)$, is defined by the limit

$$\lim_{h \to 0} \frac{f(a + h) - f(a)}{h}$$

if it exists. The existence and properties of the derivative are determined, via this definition, entirely from the existence and properties of such limits. Exactly the same ideas apply to limits, continuity and derivatives in the case of functions of two or more variables. However, in this case we do not have left- and right-sided limits, but we have the facility to approach the limit from any direction.

EXERCISES ON 3.1

1. Show that $\lim_{x \to 1} (x + 1) = 2$, using ε–δ arguments.
2. Assuming that y is held constant, evaluate

   (i) $\lim_{x \to 3} (x^2 + y^2)$ (ii) $\lim_{x \to y} \left(\dfrac{x^2 - y^2}{x - y} \right)$

3.2   Limits of functions of more than one variable

If we look carefully at the definitions and discussion in Section 3.1, we see that the only sense in which there is any restriction to a single real variable is in the definition of the intervals involved and in the definition of the modulus, as in $|f(x) - L| < \varepsilon$ for example. Now we saw in Section 2.2 that the natural extension of intervals to $\mathbb{R}^2$ are *neighbourhoods* or *discs*, and in $\mathbb{R}^3$ we have *spheres* or *balls*. This suggests the following extension of the definition of a limit to a function of two variables, $f(x, y)$. We say that the limit of $f(x, y)$ as (x, y) tends to (x_0, y_0) is L, and write this

$$\lim_{(x, y) \to (x_0, y_0)} f(x, y) = L$$

if for any number $\varepsilon > 0$ we can find a number $\delta > 0$ such that $f(x, y)$ satisfies

$$|f(x, y) - L| < \varepsilon$$

whenever (x, y) lies in the domain of f and is within a distance δ of (x_0, y_0):

$$0 < \sqrt{(x - x_0)^2 + (y - y_0)^2} < \delta$$

Note that it is not necessary that $f(x, y)$ is defined at (x_0, y_0).

This definition extends in an obvious way to a function of three variables. We write

$$\lim_{(x, y, z) \to (x_0, y_0, z_0)} f(x, y, z) = L$$

if for any number $\varepsilon > 0$ there exists a $\delta > 0$ such that

$$|f(x, y, z) - L| < \varepsilon$$

whenever (x, y, z) is in the domain of f and lies within a distance δ of (x_0, y_0, z_0), i.e.

$$0 < \sqrt{(x - x_0)^2 + (y - y_0)^2 + (z - z_0)^2} < \delta$$

Finally, we can generalize all this to $\mathbb{R}^n$ by repeating the definition with vectors $\mathbf{x}$, $\mathbf{x}_0$, and using the neighbourhood defined by

$$0 < |\mathbf{x} - \mathbf{x}_0| < \delta$$

where the modulus sign $|\cdot|$ now denotes the *norm* or length of a vector in $\mathbb{R}^n$. In the definitions given above, there is no mention of left- and right-sided limits. This is because, as defined, we can approach the above limits in any manner, along any curve, which approaches the limit point (x_0, y_0), (x_0, y_0, z_0) or $\mathbf{x}_0$. The definitions presuppose that the limit will always be the same however you approach it. If you can obtain two different values for the 'limit' by approaching along two different curves then the limit does not exist.

Before we consider some examples of limits, remind yourself about how you dealt with limits in the case of functions of a single variable. You may have seen and used the ε–δ approach for one or two examples, possibly finding it quite difficult. But you soon moved on to more direct approaches by building up a catalogue of simple standard limits which you then used in more complicated examples. In particular, you may remember that in the ε–δ approach you really have to know the limit in advance – you 'guess' it – and then use ε–δ arguments to check your guess. We do the same sort of thing in the case of limits of functions of two (or more) variables. We will work through one simple example by ε–δ methods, then build up direct methods for more complicated limits. However, in the case of two or more variables the need to evaluate limits from a number of different directions (mainly to check that the limit does not exist) means that more care is needed with such limits.

PROBLEM 2

Guess the value of the limit

$$L = \lim_{(x, y) \to (2, 1)} (2x + y)$$

Show that

$$|2x + y - L| \le 2|x - 2| + |y - 1|$$

and deduce that

$$|2x + y - L| \le 3\sqrt{(x-2)^2 + (y-1)^2}$$

Now show that we can always ensure that

$$|2x + y - L| < \varepsilon$$

by taking $\delta = \varepsilon/3$ with the notation introduced above, hence confirming the limit you guessed.

You will probably be happy with

$$\lim_{(x,\, y) \to (2,\, 1)} (2x + y) = 2 \times 2 + 1 = 5$$

but now we have to confirm this. The rest of the problem leads you through this. We find

$$|2x + y - L| = |2x + y - 5| = |2(x-2) + (y-1)|$$

(here we are setting up the $x - x_0$ and $y - y_0$)

$$\le |2(x-2)| + |y-1|$$

by the *triangle inequality*

$$|a + b| \le |a| + |b|$$

So, we have

$$|2x + y - 5| \le 2|x-2| + |y-1|$$

Now (here we are setting up ε, the distance of L from $2x + y$)

$$|x - 2| = +\sqrt{(x-2)^2} \le \sqrt{(x-2)^2 + (y-1)^2}$$

and similarly

$$|y - 1| \le \sqrt{(x-2)^2 + (y-1)^2}$$

and so we can write

$$|2x + y - 5| \le 2\sqrt{(x-2)^2 + (y-1)^2} + \sqrt{(x-2)^2 + (y-1)^2}$$

$$= 3\sqrt{(x-2)^2 + (y-1)^2}$$

as required.

If we now look back at the definition of the limit, then we see that we can always ensure that, for any positive ε,

$$|2x + y - 5| < \varepsilon$$

by choosing $\sqrt{(x-2)^2 + (y-1)^2} < \delta$, where $\delta = \varepsilon/3$, because the above inequality then gives

$$|2x + y - 5| < 3 . \frac{\varepsilon}{3} = \varepsilon$$

Hence

$$\lim_{(x,\, y) \to (2,\, 1)} (2x + y) = 5$$

as we 'guessed'.

You may have felt insecure working through this problem – this is a difficult topic. That is why we use more direct methods relying on the properties of limits and some simple basic examples. I have gone to great pains to convince you of the essential similarity between limits of several variables and limits of a single variable. It should therefore come as no surprise that the properties of such limits are also virtually identical, with obvious modifications. So I will simply list the properties without proof. If you are interested you can supply the proofs yourself by appropriately modifying the proofs for limits of a function of a single variable.

$$\lim_{(x,\, y) \to (x_0,\, y_0)} x = x_0 \qquad \lim_{(x,\, y) \to (x_0,\, y_0)} y = y_0$$

$$\lim_{(x,\, y) \to (x_0,\, y_0)} c = c \qquad \text{for any constant } c$$

If

$$\lim_{(x,\, y) \to (x_0,\, y_0)} f(x,\, y) = L \quad \text{and} \quad \lim_{(x,\, y) \to (x_0,\, y_0)} g(x,\, y) = M$$

then

$$\lim[f(x,\, y) \pm g(x,\, y)] = L \pm M$$

i.e. the limit of a sum/difference is equal to the sum/difference of the limits;

$$\lim[f(x,\, y).g(x,\, y)] = L.M$$

i.e. the limit of a product is equal to product of the limits;

$$\lim \left(\frac{f(x,\, y)}{g(x,\, y)} \right) = \frac{L}{M} \qquad \text{if } M \neq 0$$

i.e. the limit of a quotient is equal to the quotient of the limits, unless the limit in the denominator is zero.

If m and n are integers, then

$$\lim[f(x,\, y)]^{m/n} = L^{m/n}$$

provided $L^{m/n}$ is real, i.e. the limit of a power is equal to the power of the limit. This result also applies for irrational powers with appropriate care.

All these results are manifestly sensible, and they enable us to evaluate an infinite variety of limits in an obvious way, dispensing with the ε–δ arguments (which, of course, have already been used in the proofs of the above properties).

PROBLEM 3

Evaluate the limits

(i) $\displaystyle \lim_{(x,\,y)\to(-1,\,0)} \left[\frac{x+y^2}{x^2-y^2+2xy+1} \right]$

(ii) $\displaystyle \lim_{(x,\,y)\to(2,\,3)} \left[\frac{\sqrt{x^2+2y}}{\sqrt{2x^2+y^2}} \right]$

(i) $\displaystyle \lim_{(x,\,y)\to(-1,\,0)} \left[\frac{x+y^2}{x^2-y^2+2xy+1} \right] = \frac{-1+0^2}{(-1)^2-0^2+2(-1)0+1}$

$$= -\frac{1}{2}$$

Before you rush on from such an obvious example, just mentally check how this result comes about simply by applying the properties of limits and the simple results

$$\lim_{(x,\,y)\to(-1,\,0)} x = -1 \qquad\qquad \lim_{(x,\,y)\to(-1,\,0)} y = 0$$

(ii) Similarly

$$\lim_{(x,\,y)\to(2,\,3)} \left[\frac{\sqrt{x^2+2y}}{\sqrt{2x^2+y^2}} \right] = \frac{\sqrt{2^2+2\times3}}{\sqrt{2\times2^2+3^2}} = \sqrt{\frac{10}{17}}$$

PROBLEM 4

Consider the limit

$$\lim_{(x,\,y)\to(0,\,0)} \left[\frac{x-y}{x+y} \right]$$

Show that the limit does not exist by obtaining two different values by evaluating it along the *x*-axis and along the *y*-axis.

If we approach along the *x*-axis, then $y = 0$ and we have

$$\lim_{(x,\,y)\to(0,\,0)} \left[\frac{x-y}{x+y} \right] = \lim_{(x,\,y)\to(0,\,0)} \left[\frac{x}{x} \right] = 1$$

On the other hand, if we approach along the *y*-axis then $x = 0$ and we get

$$\lim_{(x,\,y)\to(0,\,0)} \left[\frac{x-y}{x+y} \right] = \lim_{(x,\,y)\to(0,\,0)} \left[\frac{-y}{y} \right] = -1$$

Thus the limit is not unique – it has at least two different values and therefore it does not exist. This example illustrates how we can show that a particular limit does not exist. If we obtain two different values for $\lim_{(x,\,y)\to(x_0,\,y_0)} f(x,\,y)$ by approaching $(x_0,\,y_0)$ by two different paths, or if the limit does not exist along some path, then $\lim_{(x,\,y)\to(x_0,\,y_0)} f(x,\,y)$ does not exist.

Note that while we can show that a limit does *not* exist using this method of evaluating it in two ways, we cannot use the method to show that the limit *does* exist – for this we have to show that it exists for *all* possible approaches – and this essentially means falling back on ε–δ arguments. Even in using the argument to show that a limit does not exist, it is often quite tricky to find ways of approaching the limit which demonstrate the fact. Here we need plenty of practice. The following problems should help.

PROBLEM 5

Repeat Problem 4 by considering the limit by approaching the origin along any straight line $y = mx$.

By taking $y = mx$ and approaching $(0, 0)$ by letting $x \to 0$ we have

$$\lim_{(x, y) \to (0, 0)} \left[\frac{x - y}{x + y} \right] = \lim_{x \to 0} \left[\frac{x - mx}{x + mx} \right]$$

$$= \frac{1 - m}{1 + m}$$

This clearly depends on the value of m and so is not unique – the limit does not exist, as we already know.

You might be tempted to think that this method of approaching the limit along a general straight line provides a possible means of showing that the limit *does* exist. Thus, if we do get the same value by approaching the limit along *any* straight line, then haven't we fulfilled the requirement to show that the limit approached is the same from any direction? Unfortunately not, as the next problem shows.

PROBLEM 6

Investigate the limit

$$\lim_{(x, y) \to (0, 0)} \left[\frac{3x^2 y}{x^4 + y^2} \right]$$

The trick $y = mx$ used above gives

$$\lim_{(x, y) \to (0, 0)} \left[\frac{3x^2 y}{x^4 + y^2} \right] = \lim_{x \to 0} \left[\frac{3mx^3}{x^4 + m^2 x^2} \right]$$

$$= \lim_{x \to 0} \left[\frac{3mx}{x^2 + m^2} \right] = 0$$

for all m. This is looking good: you may think the limit exists. However, let us now try approaching the origin along the parabola $y = mx^2$ ($m \neq 0$). This gives

$$\lim_{(x,\,y)\to(0,\,0)} \left[\frac{3x^2y}{x^4 + y^2} \right] = \lim_{x\to 0} \left[\frac{3mx^4}{x^4 + m^2x^4} \right]$$

$$= \frac{3m}{1 + m^2}$$

This depends on the value *m*: the limit is not unique, and so does not exist.

This example shows that ingenuity is sometimes required even in showing that a limit does not exist. It is not only the approach *direction* but also the approach *route* which is important. It also sets alarm bells ringing about whether you can ever be sure that a limit *does* exist – do we have to check every possible approach route to be sure a limit exists? Actually, yes – and this is what the ε–δ arguments essentially do. The most difficult limits to deal with are those that do exist, but do not lend themselves to the rules of limits we have given above.

PROBLEM 7

Investigate the limit

$$\lim_{(x,\,y)\to(0,\,0)} \left[\frac{2x(x^2 - y^2)}{x^2 + y^2} \right]$$

If you try the $y = mx$ or the $y = mx^2$ approaches you will get zero, leading to the suspicion, but not the certainty, that the limit is indeed zero. None of the rules of limits help us to confirm this, so we simply have to fall back on the ε–δ arguments. For this we need to show that given $\varepsilon > 0$ there must exist a $\delta > 0$ such that

$$\left| \frac{2x(x^2 - y^2)}{x^2 + y^2} \right| < \varepsilon$$

whenever

$$0 < \sqrt{x^2 + y^2} < \delta$$

The trick here is to express everything in the $|f(x, y) - L|$ term in terms of $\sqrt{x^2 + y^2}$. For this we use $|x| \le \sqrt{x^2 + y^2}$ and $|x^2 - y^2| \le x^2 + y^2$ to get

$$\left| \frac{2x(x^2 - y^2)}{x^2 + y^2} \right| \le \left| \frac{2\sqrt{x^2 + y^2}\,(x^2 + y^2)}{x^2 + y^2} \right| = 2\sqrt{x^2 + y^2} < 2\delta$$

So in this case we simply take $\delta = \varepsilon/2$ and thus obtain

$$\left| \frac{2x(x^2 - y^2)}{x^2 + y^2} \right| < \varepsilon$$

whenever $0 < \sqrt{x^2 + y^2} < \delta$, as required.

If I seem to have laboured this business of limits, it is because the issues of limits, continuity and differentiability are somewhat more subtle for functions of two or more variables than for the single-variable case. Going up a dimension enriches the analytical structure considerably, and we need to be clear about our basic concepts in order to appreciate this.

The extension of these ideas to functions of three or more variables is sufficiently straightforward to leave to you in the exercises. The only thing worth pointing to explicitly is the extension of the triangle inequality that is required in $\mathbb{R}^n$:

$$\left| \sum_{i=1}^{n} a_i \right| \le \sum_{i=1}^{n} |a_i|$$

EXERCISES ON 3.2

1. Use ε–δ arguments to evaluate the limit

$$\lim_{(x, y) \to (0, 1)} (x + y)$$

2. Evaluate the following limits, checking any conditions that are necessary.

(i) $\displaystyle \lim_{(x, y) \to (-1, 0)} \left[\frac{x^2 + 2xy + 3y^2}{x^4 + y^4} \right]$ (ii) $\displaystyle \lim_{(x, y) \to (1, 1)} \sqrt{\frac{x + 1}{y^2 + 2x}}$

(iii) $\displaystyle \lim_{(x, y, z) \to (1, 0, 1)} \left[\frac{x^2 + y^2 + z^2}{x + y + z} \right]$

3. Evaluate

$$\lim_{(x, y) \to (0, 0)} \left[\frac{x - y}{\sqrt{x} - \sqrt{y}} \right]$$

by multiplying top and bottom by $\sqrt{x} + \sqrt{y}$ (why can we do this?).

4. Investigate the limits

(i) $\displaystyle \lim_{(x, y) \to (0, 0)} \left[\frac{xy}{x^2 + y^2} \right]$ (ii) $\displaystyle \lim_{(x, y) \to (0, 0)} \left[\frac{xy^2}{x^2 + y^4} \right]$

3.3 Continuity

The definition of continuity of a function $f(x, y)$ at a point (x_0, y_0) is exactly analogous to the single-variable case: $f(x, y)$ is *continuous at a point* (x_0, y_0) if f is defined everywhere in a neighbourhood of (x_0, y_0) and

(i) f is defined at (x_0, y_0),

(ii) $\displaystyle \lim_{(x, y) \to (x_0, y_0)} f(x, y)$ exists,

(iii) $\displaystyle \lim_{(x, y) \to (x_0, y_0)} f(x, y) = f(x_0, y_0)$.

If any of these conditions fails then the function is *discontinuous at the point* (x_0, y_0). A function $f(x, y)$ is said to be *continuous on a set S* if it is continuous at every point of S.

Graphically, a discontinuity in a function $f(x, y)$ signals a break in the surface depicting the function – analogous to the break in the curve of a discontinuous function of a single variable. Since continuity depends crucially on the behaviour of limits, and the previous section has demonstrated how 'interesting' this can be, we can see that the study of continuity is also going to have its moments!

It follows from the properties of limits that algebraic combinations of continuous functions are also continuous, at every point at which all the functions involved are defined. For example, polynomials and rational functions are continuous everywhere in their domains. Similarly, the composition of two continuous functions is also continuous on its domain.

PROBLEM 8

Prove that

$$f(x, y) = \frac{x + 2y}{x - 1}$$

is continuous everywhere except on the line $x = 1$.

(i) Clearly, f is defined everywhere except at $x = 1$, i.e. on the line $(1, y)$.
(ii) By the properties of limits, provided $x_0 \neq 1$,

$$\lim_{(x, y)\to(x_0, y_0)} \left[\frac{x + 2y}{x - 1} \right] = \frac{x_0 + 2y_0}{x_0 - 1}$$

exists.
(iii) Provided $x_0 \neq 1$,

$$\lim_{(x, y)\to(x_0, y_0)} f(x, y) = f(x_0, y_0)$$

So $f(x, y)$ is continuous everywhere except on the line $x = 1$.

PROBLEM 9

Prove that

$$f(x, y) = \begin{cases} \dfrac{xy}{x^2 + y^2} & (x, y) \neq (0, 0) \\ 0 & (x, y) = (0, 0) \end{cases}$$

is not continuous at the origin.

(i) $f(x, y)$ is clearly defined at $(x, y) = (0, 0)$. However,
(ii) from Exercise 3.2.4

$$\lim_{(x, y)\to(0, 0)} \left[\frac{xy}{x^2 + y^2} \right]$$

is not defined, and so condition (ii) is violated and $f(x, y)$ is not continuous at the origin.

The above definitions and properties of continuity, depending essentially on the properties of limits, are easily extended to functions of more than two variables.

Note that for functions of two (or more) variables it is possible to define continuity in each variable separately in an obvious way. For example, we say $f(x, y)$ is continuous in x at (x_0, y_0) if the limit $\lim_{x \to x_0} f(x, y_0)$ exists and is equal to $f(x_0, y_0)$.

As you would expect, a continuous function of several variables is continuous in each of its variables separately. However, as Exercise 3.3.3 shows, the converse is not necessarily true – a function may be continuous in each of its variables separately, and yet still be discontinuous as a function of several variables.

EXERCISES ON 3.3

1. Examine the continuity or otherwise of the following functions:

(i)
$$f(x, y) = \begin{cases} \dfrac{3xy + y^3}{x^2 + y^2} & (x, y) \neq (0, 0) \\ 0 & (x, y) = (0, 0) \end{cases}$$

(ii)
$$f(x, y) = \begin{cases} \dfrac{xy^2}{x^2 + y^2} & (x, y) \neq (0, 0) \\ 0 & (x, y) = (0, 0) \end{cases}$$

2. Prove that
$$\lim_{(x, y) \to (0, 0)} \left[\frac{xy(y^2 - x^2)}{x^2 + y^2} \right] = 0$$

Discuss the continuity of the function
$$f(x, y) = \begin{cases} \dfrac{xy(y^2 - x^2)}{x^2 + y^2} & (x, y) \neq (0, 0) \\ 1 & (x, y) = (0, 0) \end{cases}$$

3. Referring to Problem 9, show that the function defined there *is* continuous in x and y *separately*.

3.4 Partial derivatives as limits

We will be looking at the techniques of partial differentiation in Chapter 4, but here we will introduce the ideas as applications of limits. In Section 3.1 you reminded yourself about the limit definition of the derivative in the case of a function of a single variable. Thus the *derivative* of $y = f(x)$ at the point (x, y) is defined by

$$\frac{df(x)}{dx} = \lim_{h \to 0} \left(\frac{f(x + h) - f(x)}{h} \right)$$

provided this limit exists. If it does, then we say $f(x)$ is *differentiable* at (x, y). Graphically, this derivative is the *slope* of the curve $y = f(x)$, at the point (x, y).

A straightforward extension of this definition defines the *first-order partial derivative of $z = f(x, y)$ with respect to x:*

$$\frac{\partial z}{\partial x} = \frac{\partial f(x, y)}{\partial x} = \lim_{h \to 0} \left(\frac{f(x + h, y) - f(x, y)}{h} \right)$$

PROBLEM 10

Using the above definition obtain the partial derivative of $z = x^2 y + xy$ with respect to x.

In this case

$$f(x, y) = x^2 y + xy$$

$$f(x + h, y) = (x + h)^2 y + (x + h)y$$

So

$$\frac{\partial f}{\partial x} = \lim_{h \to 0} \left(\frac{(x + h)^2 y + (x + h)y - x^2 y - xy}{h} \right)$$

$$= \lim_{h \to 0} \left(\frac{2hxy + h^2 y + hy}{h} \right)$$

$$= \lim_{h \to 0} \left(2xy + hy + y \right)$$

$$= 2xy + y$$

Note that this is exactly what we would get if we differentiated $f(x, y)$ as a function of x, treating y as if it were a constant. Indeed, this is the 'practical' way of calculating partial derivatives. It is entirely different from 'totally' differentiating $f(x, y)$ with respect to x, treating y as a function of x – this would give

$$\frac{d}{dx}(x^2 y + xy) = 2xy + x^2 \frac{dy}{dx} + y + x \frac{dy}{dx}$$

which, of course, reduces to $2xy + y$ if y is a constant, since then $dy/dx = 0$.

If you review the above argument for finding $\partial z/\partial x$, you will see that we can equally well consider the derivative of $z = f(x, y)$ with respect to y, regarding x as a fixed quantity. This is defined by

$$\frac{\partial z}{\partial y} = \frac{\partial f}{\partial y} = \lim_{k \to 0} \left(\frac{f(x, y + k) - f(x, y)}{k} \right)$$

and is called the *first-order partial derivative of $f(x, y)$ with respect to y*.

PROBLEM 11

Find $\partial f/\partial y$ for $f(x, y) = x^2 y + xy$.

$$\frac{\partial f}{\partial y} = \lim_{k \to 0} \left(\frac{x^2(y+k) + x(y+k) - x^2 y - xy}{k} \right)$$

$$= \lim_{k \to 0} \left(\frac{x^2 k + xk}{k} \right)$$

$$= \lim_{k \to 0} (x^2 + x) = x^2 + x$$

Again, this is what we get if we differentiate $f(x, y)$ with respect to y, keeping x constant:

$$\frac{\partial}{\partial y}(x^2 y + xy) = x^2 + x$$

EXERCISES ON 3.4

1. Using the limit definition obtain the two partial derivatives of each of the functions

   (i) xy (ii) $x + 2y$ (iii) $3x^2 y - 2xy^2$ (iv) $\cos(xy)$ (v) e^{xy}

2. Verify your answers in Exercise 1 by differentiating with respect to $x(y)$ keeping $y(x)$ constant. In cases (i), (iv), (v) spot the short-cut.

3.5 The rules of partial differentiation derived from the properties of limits

The rules of partial differentiation may be derived from the properties of limits, in exactly the same way that they are in the case of functions of a single variable. Thus, we can treat the rules of differentiation as applications of the properties of limits. You may have forgotten the single-variable proofs, but the following problems should soon refresh your memory. The general approach is always the same – apply the limit definition to the combined functions and decompose the result into limit expressions for the component functions, using the properties of limits given in Section 3.2.

● *The sum rule* —————————————————————————

PROBLEM 12

Show that if $f(x, y)$, $g(x, y)$ are two differentiable functions then

$$\frac{\partial(f+g)}{\partial x} = \frac{\partial f}{\partial x} + \frac{\partial g}{\partial x}$$

and similarly for the y-derivative.

This follows simply from the fact that the limit of a sum is the sum of the limits, so

$$\frac{\partial (f+g)}{\partial x} = \lim_{h \to 0} \left(\frac{(f+g)(x+h, y) - (f+g)(x, y)}{h} \right)$$

$$= \lim_{h \to 0} \left(\frac{f(x+h, y) + g(x+h, y) - f(x, y) - g(x, y)}{h} \right)$$

$$= \lim_{h \to 0} \left(\frac{f(x+h, y) - f(x, y)}{h} + \frac{g(x+h, y) - g(x, y)}{h} \right)$$

$$= \lim_{h \to 0} \left(\frac{f(x+h, y) - f(x, y)}{h} \right) + \lim_{h \to 0} \left(\frac{g(x+h, y) - g(x, y)}{h} \right)$$

$$= \frac{\partial f}{\partial x} + \frac{\partial g}{\partial x}$$

Similarly for the *y*-derivative.

● *The product rule* ───────────────────────

PROBLEM 13

Starting from

$$\frac{\partial (fg)}{\partial x} = \lim_{h \to 0} \left(\frac{f(x+h, y)g(x+h, y) - f(x, y)g(x, y)}{h} \right)$$

add and subtract $f(x + h, y)g(x, y)$ to prove the product rule

$$\frac{\partial (fg)}{\partial x} = f \frac{\partial g}{\partial x} + g \frac{\partial f}{\partial x}$$

We have

$$\frac{\partial (fg)}{\partial x} = \lim_{h \to 0} \left(\frac{f(x+h, y)g(x+h, y) - f(x, y)g(x, y)}{h} \right)$$

$$= \lim_{h \to 0} \left(\frac{f(x+h, y)g(x+h, y) + f(x+h, y)g(x, y) - f(x+h, y)g(x, y) - f(x, y)g(x, y)}{h} \right)$$

$$= \lim_{h \to 0} \left(f(x+h, y) \left(\frac{g(x+h, y) - g(x, y)}{h} \right) \right) + \lim_{h \to 0} \left(g(x, y) \left(\frac{f(x+h, y) - f(x, y)}{h} \right) \right)$$

$$= f(x, y) \frac{\partial g(x, y)}{\partial x} + g(x, y) \frac{\partial f(x, y)}{\partial x}$$

Similarly for the *y*-derivative :

$$\frac{\partial (fg)}{\partial y} = f(x, y) \frac{\partial g(x, y)}{\partial y} + g(x, y) \frac{\partial f(x, y)}{\partial y}$$

● *The quotient rule* ───────────────────────

In this case I will just provide the outline, and you can supply the details as an exercise. Provided $g(x, y) \neq 0$ in the region of interest, we have

$$\frac{\partial(f/g)}{\partial x} = \lim_{h \to 0} \left(\frac{\frac{f(x+h, y)}{g(x+h, y)} - \frac{f(x, y)}{g(x, y)}}{h} \right)$$

$$= \lim_{h \to 0} \left(\frac{f(x+h, y)g(x, y) - f(x, y)g(x+h, y)}{hg(x+h, y)g(x, y)} \right)$$

$$= \lim_{h \to 0} \left(\frac{\frac{(f(x+h, y) - f(x, y))}{h} g(x, y) - f(x, y)\frac{(g(x+h, y) - g(x, y))}{h}}{g(x+h, y)g(x, y)} \right)$$

on adding and subtracting $f(x, y)g(x, y)$ (we get a lot for nothing in mathematics!)

$$= \frac{g\dfrac{\partial f}{\partial x} - f\dfrac{\partial g}{\partial x}}{g^2}$$

In the above derivations we have not worried too much about analytical details or the various conditions which have to be satisfied. This is, as always, the inevitable compromise between the need for rigour and the need to progress rapidly through the methodology.

EXERCISES ON 3.5

1. Prove the function of a function rule

$$\frac{\partial f(g(x), y)}{\partial x} = \frac{\partial g}{\partial x}\frac{\partial f(g, y)}{\partial g}$$

Note that care is needed here in realizing that we are only talking about the function of a function in one of the variables at a time. State the corresponding rule for the y-derivative.

2. Evaluate the x- and y-derivatives of the following functions using the rules of differentiation:

   (i) $2x + 3y$ (ii) x^2y (iii) $\dfrac{xy}{x+y}$ (iv) $e^{x^2}y^2$

FURTHER EXERCISES

1. Evaluate the following limits using ε–δ arguments:

   (i) $\lim\limits_{x \to 2} (2x - 3)$ (ii) $\lim\limits_{x \to 0} \left(\dfrac{x^2}{x} \right)$ (iii) $\lim\limits_{x \to 1} \left(\dfrac{x^2 - 1}{x - 1} \right)$

2. Assuming that x is a non-zero constant, evaluate

   (i) $\lim\limits_{y \to 2} \left(\dfrac{x^2 - y^2}{x + y} \right)$ (ii) $\lim\limits_{y \to 0} \left(\dfrac{x - e^y}{x^2 + \cos y} \right)$ (iii) $\lim\limits_{y \to 0} \left(\dfrac{\sin(xy)}{y} \right)$

   (iv) $\lim\limits_{y \to 1} \left(\sqrt{\dfrac{x^2 - y}{x - y}} \right)$ (v) $\lim\limits_{y \to -2} \left(\dfrac{x^2 + 2yx + y^2}{x - 2} \right)$

   State any results used, and the conditions associated with them.

3. Use ε–δ arguments to evaluate the following limits:

(i) $\displaystyle\lim_{(x,\,y)\to(0,\,1)} (3x + 4y)$

(ii) $\displaystyle\lim_{(x,\,y)\to(1,\,0)} \left(\frac{x^2 - y^2}{x + y}\right)$

(iii) $\displaystyle\lim_{(x,\,y)\to(2,\,1)} (ax + by)$

(iv) $\displaystyle\lim_{(x,\,y)\to(1,\,1)} \left(\frac{x}{3y}\right)$

(v) $\displaystyle\lim_{(x,\,y)\to(0,\,1)} (x^2 + 2y^2 + 1)$

4. Evaluate the following limits:

(i) $\displaystyle\lim_{(x,\,y)\to(1,\,1)} \left(\frac{x^2 + 3y^2}{2x^3 + 4xy + y^2}\right)$

(ii) $\displaystyle\lim_{(x,\,y)\to(0,\,0)} \left(\frac{x^3y + yx + 4}{xy^2 - 3xy + 7}\right)$

(iii) $\displaystyle\lim_{(x,\,y)\to(1,\,-1)} \left(\frac{x + y}{x - y}\right)$

(iv) $\displaystyle\lim_{(x,\,y)\to(0,\,1)} \left(\frac{\sin(x + y)}{(x + y)}\right)$

(v) $\displaystyle\lim_{(x,\,y)\to(0,\,0)} \left(\frac{e^{-xy}}{\cos(x + y)}\right)$

(vi) $\displaystyle\lim_{(x,\,y)\to(1,\,1)} \left(\frac{\sin(x - y)}{\cos(x - y)}\right)$

(vii) $\displaystyle\lim_{(x,\,y)\to(2,\,1)} \ln(x^2y + 5x + 6y - 1)$

(viii) $\displaystyle\lim_{(x,\,y)\to(0,\,0)} \left(\frac{1 + xy}{1 - xy}\right)$

5. Show that each of the following limits does not exist:

(i) $\displaystyle\lim_{(x,\,y)\to(0,\,0)} \left(\frac{x + y}{x - y}\right)$

(ii) $\displaystyle\lim_{(x,\,y)\to(0,\,0)} \left(\frac{xy + 3x^3}{x^2 + y^2}\right)$

(iii) $\displaystyle\lim_{(x,\,y)\to(0,\,0)} \left(\frac{x^2 - y^2}{x^2 + y^2}\right)$

(iv) $\displaystyle\lim_{(x,\,y)\to(0,\,0)} \left(\frac{x - y}{x^2 + y^2}\right)$

(v) $\displaystyle\lim_{(x,\,y)\to(0,\,0)} \left(\frac{xy}{2x^2 + 3y^2}\right)$

(vi) $\displaystyle\lim_{(x,\,y)\to(0,\,0)} \left(\frac{(x^2 + y^2)^2}{x^4 + y^4}\right)$

6. Evaluate the following limits:

(i) $\displaystyle\lim_{(x,\,y,\,z)\to(2,\,1,\,1)} \ln(x + 3yz - 4y^2z)$

(ii) $\displaystyle\lim_{(x,\,y,\,z)\to(2,\,-1,\,2)} \left(\frac{xz^2}{\sqrt{x^2 + y^2 + z^2}}\right)$

(iii) $\displaystyle\lim_{(x,\,y,\,z)\to(0,\,0,\,0)} \left(\frac{\sin(x^2 + y^2 + z^2)}{\sqrt{x^2 + y^2 + z^2}}\right)$

7. Show that the following limits do not exist:

(i) $\displaystyle\lim_{(x,\,y,\,z)\to(0,\,0,\,0)} \left(\frac{2xy + 3xz + yz}{x^2 + y^2 + z^2}\right)$

(ii) $\displaystyle\lim_{(x,\,y,\,z)\to(0,\,0,\,0)} \left(\frac{xyz}{x^3 + y^3 + z^3}\right)$

8. Determine whether or not the following functions are continuous at the points indicated:

(i) $f(x, y) = x^2 + 2y^2$ at $(1, 1)$

(ii) $f(x, y) = \dfrac{x^2 + y^2}{x - y}$ at $(0, 1)$

(iii) $f(x, y) = \dfrac{e^x - e^y}{x - y}$ at $(2, 2)$

(iv) $f(x, y) = \dfrac{\sin(x^2 + y^2)}{xy}$ at $(1, 1)$

(v) $f(x, y) = \dfrac{x^3 - 2xy^2 + y - 1}{y^3 + 2x^2y - y + 2}$ at $(1, 0)$

(vi) $f(x, y) = \dfrac{x^2 + y^2}{x^2 - 4xy + 4y^2}$ at $(2, 1)$

9. Define $f(0, 0)$ in a way such that

$$f(x, y) = xy\,\frac{x^2 - y^2}{x^2 + y^2}$$

extends to be continuous at the origin. (*Hint:* cf. Exercise 3.3.2.)

10. Determine whether or not the following functions are continuous:

(i)
$$f(x, y) = \begin{cases} \dfrac{xy^2}{x^2 + y^4} & (x, y) \neq (0, 0) \\ 0 & (x, y) = (0, 0) \end{cases}$$

(ii)
$$f(x, y) = \begin{cases} \dfrac{\sin(x^2 + y^2)}{x^2 + y^2} & (x, y) \neq (0, 0) \\ 1 & (x, y) = (0, 0) \end{cases}$$

(iii)
$$f(x, y) = \begin{cases} \dfrac{2xy}{x^2 + y^2} & (x, y) \neq (0, 0) \\ 0 & (x, y) = (0, 0) \end{cases}$$

(iv)
$$f(x, y) = \begin{cases} \dfrac{x^2y^2}{x^2 + y^2} & (x, y) \neq (0, 0) \\ 0 & (x, y) = (0, 0) \end{cases}$$

(*Hint:* use polar coordinates.)

11. In each case describe the maximum region over which the given function $f(x, y)$ is continuous:

(i) $f(x, y) = \sqrt{y - x}$

(ii) $f(x, y) = \dfrac{2x^3 + xy^6 - 3x^4}{x^3 - y^3}$

(iii) $f(x, y) = \ln(2x - 3y + 3)$

(iv) $f(x, y) = e^{xy}$

(v) $f(x, y) = \dfrac{x}{\sqrt{1 - (x/2)^2 - y^2}}$

(vi) $f(x, y, z) = \dfrac{xyz - y^3 + xz^2}{x + 2y - z + 2}$

(vii) $f(x, y, z) = x - \sin^{-1}(z^2 + y)$

(viii) $f(x, y, z) = \dfrac{1}{\sqrt{1 - x^2 - y^2 - z^2}}$

12. Find a constant c such that the function

$$f(x, y) = \begin{cases} \dfrac{x^2 y^2}{\sqrt{x^2 + y^2}} & (x, y) \neq (0, 0) \\ c & (x, y) = (0, 0) \end{cases}$$

is continuous at the origin.

13. Define $f(0, 0)$ such that f extends to be continuous at the origin:

(i) $f(x, y) = \ln\left(\dfrac{2x^2 - 3x^2 y^2 + 2y^2}{x^2 + y^2}\right)$ (ii) $f(x, y) = \dfrac{xy^2}{x^2 + y^2}$

14. Find a function $g(x)$ such that the function

$$f(x, y) = \begin{cases} \dfrac{x^2 - y^2}{x - y} & x \neq y \\ g(x) & x = y \end{cases}$$

is continuous at every point of $\mathbb{R}^3$.

15. Evaluate the partial derivatives with respect to x and y of the following functions by means of the limit definition:

(i) $f(x, y) = 2x + y$ (ii) $f(x, y) = x^2 + 3y^2$

(iii) $f(x, y) = \dfrac{1}{xy}$ (iv) $f(x, y) = \sin(xy)$

(v) $f(x, y) = e^{x+y}$

16. If

$$f(x, y) = \begin{cases} \dfrac{x^2 y^2}{x^4 + y^4} & (x, y) \neq (0, 0) \\ 0 & (x, y) = (0, 0) \end{cases}$$

(i) show that $\partial f/\partial x$ and $\partial f/\partial y$ both exist at $(0, 0)$, and give their values at $(0, 0)$
(ii) show that $\lim_{(x, y) \to (0, 0)} f(x, y)$ does not exist.

17. Essay or discussion topics:
   (i) The fate of single-sided limits in going from functions of one variable to two variables.
   (ii) ε–δ proofs of the properties of limits.
   (iii) The use of counter-examples in showing that limits do not exist.
   (iv) The circumstances under which continuity breaks down for functions of two or more variables.
   (v) Continuity away from the origin.
   (vi) Partial derivatives as limits.
   (vii) Limits, continuity and derivatives for functions of three or more variables.

4 • Differentiation of Functions of More Than One Variable

4.1 Partial derivatives and their properties

In Section 3.4 we saw how, as a direct extension of the limit definition of the ordinary derivative, we can define the first-order partial derivative of $f(x, y)$ with respect to x as

$$f_x = \frac{\partial f(x, y)}{\partial x} = \lim_{h \to 0} \left(\frac{f(x + h, y) - f(x, y)}{h} \right)$$

and similarly for the corresponding partial derivative with respect to y

$$f_y = \frac{\partial f(x, y)}{\partial y} = \lim_{k \to 0} \left(\frac{f(x, y + k) - f(x, y)}{k} \right)$$

To see the geometrical significance of $\partial f / \partial x$, consider the surface generated by the function $z = f(x, y)$ (Fig. 1.2). Fixing the value of y defines a plane parallel to the xz-plane, through the point (x, y). The intersection of this plane with the surface defines a curve on the surface. The partial derivative $\partial f / \partial x$ gives the slope of this curve at the point (x, y, z).

Similarly, $\partial f / \partial y$ gives the gradient of the curve generated by the intersection of the surface $z = f(x, y)$ with the plane parallel to the yz-plane through the point (x, y, z).

As remarked in Chapter 3, $\partial f / \partial x$, if it exists, is obtained by differentiating $f(x, y)$ with respect to x, regarding y as a constant. Similarly, in evaluating $\partial f / \partial y$ we regard x as a constant. Apart from this, the rules of partial differentiation are the same as for ordinary differentiation. We proved some of the rules in Section 3.4, as applications of the properties of limits; here we will just summarize them:

$$\frac{\partial}{\partial x} (f(x, y) \pm g(x, y)) = \frac{\partial f(x, y)}{\partial x} \pm \frac{\partial g(x, y)}{\partial x}$$

$$\frac{\partial}{\partial x} (f(x, y) g(x, y)) = f(x, y) \frac{\partial g(x, y)}{\partial x} + \frac{\partial f(x, y)}{\partial x} g(x, y)$$

$$\frac{\partial}{\partial x} \left(\frac{f(x, y)}{g(x, y)} \right) = \frac{g(x, y) \dfrac{\partial f(x, y)}{\partial x} - f(x, y) \dfrac{\partial g(x, y)}{\partial x}}{(g(x, y))^2} \qquad g(x, y) \neq 0$$

$$\frac{\partial f(g(x), y)}{\partial x} = \frac{\partial g}{\partial x} \frac{\partial f(g, y)}{\partial g}$$

with similar results for the partial derivatives with respect to y.

Particularly if *f* is a function of three or more variables, the statement that $\partial f/\partial x$, for example, is *f* differentiated with respect to *x* with *y* held constant is not always unambiguous. It is therefore sometimes convenient to use the notation

$$\left(\frac{\partial f}{\partial x}\right)_y$$

However, this clumsy notation can usually be avoided by being clear about the context of the derivative.

EXERCISE ON 4.1

Find f_x, f_y for the following functions:

(i) $f(x, y) = \dfrac{xy \tan(x + y)}{x + y}$ (ii) $e^{3x+\cos(xy)}$ (iii) $\dfrac{x^3 + y}{xy^2 + 1}$

4.2 Higher-order derivatives

f_x, f_y are in general still both functions of *x* and *y* and so may be differentiated again with respect to either variable. We write, for example,

$$\frac{\partial}{\partial x}\left(\frac{\partial f}{\partial x}\right) = \frac{\partial^2 f}{\partial x^2} \equiv f_{xx}$$

This is called the *second-order partial derivative with respect to x*. Similarly,

$$\frac{\partial}{\partial y}\left(\frac{\partial f}{\partial y}\right) = \frac{\partial^2 f}{\partial y^2} \equiv f_{yy}$$

We can also calculate *mixed derivatives*,

$$\frac{\partial}{\partial y}\left(\frac{\partial f}{\partial x}\right) = \frac{\partial^2 f}{\partial y \partial x} \equiv f_{xy}$$

and

$$\frac{\partial}{\partial x}\left(\frac{\partial f}{\partial y}\right) = \frac{\partial^2 f}{\partial x \partial y} \equiv f_{yx}$$

(note carefully the order of *x*, *y* in these results).

Now, it might not occur to you, but trying a few examples will soon persuade you that the mixed derivatives *may* be equal:

$$\frac{\partial^2 f}{\partial x \partial y} = \frac{\partial^2 f}{\partial y \partial x}$$

or

$$f_{xy} = f_{yx}$$

In fact, given certain conditions, this is indeed the case. The precise result is as follows. If *f* and f_x, f_y, f_{xy} and f_{yx} are all continuous on an open set *S*, then on *S* the mixed derivatives are equal – i.e.

$$\frac{\partial^2 f}{\partial x \partial y} = \frac{\partial^2 f}{\partial y \partial x}$$

This result was first stated by Euler in a hydrodynamics paper published in 1734. The proof uses the mean value theorem (see Chapter 5) and may be found in Thomas and Finney (1996), for example.

PROBLEM I

> Find all second-order derivatives of the function
>
> $$f(x, y) = x^2 e^{x+y} + xy^3$$

$$f_x = 2xe^{x+y} + x^2 e^{x+y} + y^3$$
$$f_y = x^2 e^{x+y} + 3xy^2$$
$$f_{xx} = x^2 e^{x+y} + 4xe^{x+y} + 2e^{x+y}$$
$$f_{xy} = 2xe^{x+y} + x^2 e^{x+y} + 3y^2 = f_{yx}$$
$$f_{yy} = x^2 e^{x+y} + 6xy$$

There is, of course, no reason why we should stop at second-order derivatives – we can keep on differentiating to our hearts' content, provided the appropriate conditions hold. Apart from this it is simply a matter of notation, which again is fairly self-evident. For example,

$$\frac{\partial^3 f}{\partial y^2 \partial x} \equiv \frac{\partial}{\partial y}\left(\frac{\partial}{\partial y}\left(\frac{\partial f}{\partial x}\right)\right) \equiv f_{xyy}$$

$$\frac{\partial^4 f}{\partial x^2 \partial y^2} \equiv \frac{\partial}{\partial x}\left(\frac{\partial}{\partial x}\left(\frac{\partial}{\partial y}\left(\frac{\partial f}{\partial y}\right)\right)\right) \equiv f_{yyxx}$$

and so on.

And again, provided all appropriate derivatives are continuous on an open set, then everywhere on that open set the order of the differentiation is irrelevant – for example,

$$f_{xyx} = f_{xxy} = f_{yxx}$$

In practice, there is no great call for derivatives higher than second order – as you might suspect from the number of times you have had to deal with ordinary derivatives of greater than second order. A notable exception occurs in the theory of elasticity, where the *biharmonic equation* contains fourth-order derivatives, but that is perhaps stretching things a bit.

EXERCISES ON 4.2

1. Find all first- and second-order partial derivatives, checking the equality of the mixed derivatives, for the following functions:
   (i) $x^3 y^2 + 4xy^4$ (ii) $\ln(x^2 + y^2)$ (iii) $e^{xy} \cos(x + y)$
2. Show that $f(x, y) = \ln(x^2 + y^2)$ satisfies the *partial differential equation*

   $$\frac{\partial^2 f}{\partial x^2} + \frac{\partial^2 f}{\partial y^2} = 0$$

   This is called the *Laplace equation* in two-dimensional rectangular coordinates.

It is very important in fluid mechanics, electromagnetism and many other areas of science and engineering, as well as being a key equation in pure mathematics.

4.3 Differentiation of functions of more than two variables

I wonder if, like me, you find diagrams such as Fig. 1.2 less than enlightening? Certainly, such geometrical representations do not appeal to everyone. But they are clearly important, and they certainly give meaning to the idea of a derivative of a single variable, for example. But we should not forget that they are only diagrammatic illustrations of concepts which are not necessarily intrinsically geometric. For example, the slope of a curve is basically a geometric representation of a means of measuring the rate at which something is increasing. Similarly, a derivative of a function of two variables tells us how fast the function is increasing in a particular direction – it does not *have* to be viewed as the slope of a particular surface, in a particular direction. And, as soon as we start to talk about functions of *more than two* variables, the direct usefulness of the geometric representation starts to decline rapidly. We can just about manage perspective pictures of surfaces in three-dimensional space, but beyond that it becomes impossible even to picture the functions. This is not to say that geometric intuition and argument become useless in higher dimensions (you have already seen how progress in three dimensions was facilitated by reference back to one dimension), rather that this now works by analogy and not direct appeal to the senses.

So what do we do when faced with functions of more than two variables – functions we cannot 'sketch' on a two-dimensional piece of paper? Well, as always in mathematics, we look for natural ways of extending the symbolism to 'perpetuate the present pattern'. For example, we extend from three-dimensional vector spaces with 3×1 column matrix representation to n-dimensional vector spaces simply by extending the column matrices to $n \times 1$ vectors. We then find that many things we did in three dimensions extend naturally to n dimensions. For example, the 'length' of a three-dimensional vector $(x_1, x_2, x_3)^T$ is defined as

$$\sqrt{x_1^2 + x_2^2 + x_3^2}$$

while the 'length' of an n-dimensional vector $(x_1, x_2, ..., x_n)^T$ is defined to be

$$\sqrt{x_1^2 + x_2^2 + ... + x_n^2}$$

So, we now look back over the previous sections to see what 'natural' extensions can be made when we go to functions of three or more variables.

Thus, let

$$f(x_1, x_2, ..., x_n)$$

be a function of n independent variables $x_1, x_2, ..., x_n$. Then the definition of the derivative in Section 4.1 extends quite naturally to

$$\frac{\partial f}{\partial x_1} = \lim_{h \to 0} \left(\frac{f(x_1 + h, x_2, ..., x_n) - f(x_1, x_2, ..., x_n)}{h} \right)$$

In a way, the extra $n - 2$ variables make little difference at all – except we can of course now differentiate with respect to them. For example,

$$\frac{\partial f}{\partial x_n} = \lim_{k \to 0} \left(\frac{f(x_1, x_2, \ldots, x_n + k) - f(x_1, x_2, \ldots, x_n)}{h} \right)$$

In practical terms the extension is also clear. To differentiate f with respect to a given variable x_i, we simply differentiate as if all other variables are fixed constants.

PROBLEM 2

Find all second-order partial derivatives of

$$f(x, y, z) = x^4 y^3 z^2$$

$$\frac{\partial f}{\partial x} = 4x^3 y^3 z^2$$

$$\frac{\partial^2 f}{\partial x^2} = 12x^2 y^3 z^2$$

$$\frac{\partial f}{\partial y} = 3x^4 y^2 z^2$$

$$\frac{\partial^2 f}{\partial y^2} = 6x^4 y z^2$$

$$\frac{\partial f}{\partial z} = 2x^4 y^3 z$$

$$\frac{\partial^2 f}{\partial z^2} = 2x^4 y^3$$

$$\frac{\partial^2 f}{\partial x \partial y} = \frac{\partial^2 f}{\partial y \partial x} = 12x^3 y^2 z^2$$

$$\frac{\partial^2 f}{\partial x \partial z} = \frac{\partial^2 f}{\partial z \partial x} = 8x^3 y^3 z$$

$$\frac{\partial^2 f}{\partial y \partial z} = \frac{\partial^2 f}{\partial z \partial y} = 6x^4 y^2 z$$

It should now be clear that partial differentiation is little more demanding than ordinary differentiation, in principle. But then it does not really give us much anyway. When we partially differentiate $f(x, y)$ with respect to x, for example, all it gives us is the rate of change of f with respect to x, assuming that y is kept constant. This is clearly not dealing with $f(x, y)$ as a function of two *variables*. Obviously, the much more interesting question is how $f(x, y)$ varies as *both x and y* vary. This is the subject of Section 4.5.

EXERCISES ON 4.3

1. Find all third-order derivatives of the function $f(x, y, z) = (x + \sin z)e^{x-y}$.

2. A function of n variables is defined by

$$f(x_1, x_2, \ldots, x_n) = x_1^n x_2^n \ldots x_n^n$$

where n is an integer. Evaluate the expression

$$\sum_{i=1}^{n} x_i \frac{\partial f}{\partial x_i}$$

4.4 Partial differential equations

In the case of functions of a single variable, any equation containing an ordinary derivative is called an *ordinary differential equation* (*see* Pearson, 1996; Cox, 1996). Analogously, in the case of functions of two variables, any equation containing one or more partial derivatives is called a *partial differential equation* (PDE). Most of the definitions in PDEs are exactly as you would expect by analogy with ordinary differential equations. Specifically, if $z = f(x, y)$ is a continuous function of x and y, with sufficiently high-order derivatives, then any equation containing up to any one of the *n*th-order partial derivatives is called an *nth-order PDE for the dependent function z, in the independent variables x, y.*

PROBLEM 3

> Find the order of the following PDEs:
> (i) $xf_x + yf_y = x^2$ (ii) $z_{xy} = 0$
>
> (iii) $\dfrac{\partial^4 \phi}{\partial x^4} + \dfrac{\partial^4 \phi}{\partial y^4} + \dfrac{\partial^4 \phi}{\partial z^4} = 0$

Part (i) contains only first-order derivatives and so is a first-order PDE. Similarly, (ii) is second order and (iii) is fourth order.

Note that nothing is said about the dependent functions f, z, ϕ. You might have assumed that f and z are functions of x and y and that ϕ is a function of x, y, z. This would be presumptuous, however. Each of them may be functions of any number of variables – these do not necessarily have to appear as derivatives or coefficients in the PDEs.

The question obviously arises about the solution of PDEs. Even the simplest require care:

PROBLEM 4

> Solve the following PDEs:
> (i) $\dfrac{\partial z\,(x,\,y)}{\partial x} = 0$ (ii) $\dfrac{\partial z\,(x,\,y)}{\partial y} = x$
>
> (iii) $\dfrac{\partial^2 z\,(x,\,y)}{\partial x \partial y} = 0$ (iv) $\dfrac{\partial^2 z\,(x,\,y)}{\partial x^2} = 0$

(i) When z_x is evaluated it is assumed that y remains constant. So, for *any* function of y, say $g(y)$, we have

$$\frac{\partial g(y)}{\partial x} = 0$$

So one solution of $z_x = 0$ is certainly $z = g(y)$, an arbitrary function of y alone. In fact, this is the *general solution*. Arbitrary *functions* in the solution of PDEs replace arbitrary constants of integration in ordinary differential equations.

(ii) To solve $z_y = x$ we must integrate the x, with respect to y, but treating x as a constant. But at the same time we must add an arbitrary function of x to z on integration. Thus

$$z = \int x\,dy + g(x)$$

$$= xy + g(x)$$

with $g(x)$ an arbitrary function.

(iii) We have

$$\frac{\partial^2 z}{\partial x \partial y} \equiv \frac{\partial}{\partial x}\left(\frac{\partial z}{\partial y}\right) = 0$$

so, integrating once,

$$\frac{\partial z}{\partial y} = f(y)$$

where $f(y)$ is an arbitrary function of y. Integrating again,

$$z = \int f(y)\,dy + g(x)$$

So the general solution is

$$z = h(y) + g(x)$$

for arbitrary functions h and g. Note that for this second-order PDE we get *two* arbitrary functions in the general solution.

(iv) Integrating once gives

$$\frac{\partial z}{\partial x} = f(y)$$

Integrating again gives

$$z = xf(y) + g(y)$$

Again, this contains two arbitrary functions.

Obviously the solution of more complicated PDEs soon becomes very difficult, and this is not the place to go deeply into the subject. However, there is an important method of solving some PDEs which is instructive, and which is not too difficult to appreciate – this is *separation of variables*. We will look at this via an example.

PROBLEM 5

> Consider the PDE
>
> $$\frac{\partial^2 z}{\partial x^2} + \frac{\partial^2 z}{\partial y^2} = 0$$
>
> for the function $z(x, y)$. Assume a solution of the separable form
>
> $$z = f(x)g(y)$$
>
> where $f(x)$ is a function of x alone, $g(y)$ a function of y alone. Show that f and g must satisfy ordinary differential equations.

Essentially, this trick sometimes allows us to replace partial derivatives by ordinary derivatives. Thus

$$\frac{\partial^2 z}{\partial x^2} = \frac{d^2 f(x)}{dx^2} g(y)$$

$$\frac{\partial^2 z}{\partial y^2} = f(x) \frac{d^2 g(y)}{dy^2}$$

So the PDE becomes

$$\frac{d^2 f(x)}{dx^2} g(y) + f(x) \frac{d^2 g(y)}{dy^2} = 0$$

We can rearrange this as

$$\frac{1}{f(x)} \frac{d^2 f(x)}{dx^2} = -\frac{1}{g(y)} \frac{d^2 g(y)}{dy^2}$$

Now comes the pretty bit. The left-hand side is clearly a function of x only, and the right-hand side is a function of y only. But x and y are *independent variables.* There can be no relation between them. So the only way the above equation can be true is if both sides are actually independent of x and y, and so they can only be *constant.* So we must have

$$\frac{1}{f(x)} \frac{d^2 f(x)}{dx^2} = -\frac{1}{g(y)} \frac{d^2 g(y)}{dy^2} = \lambda$$

where λ is a constant. This yields the ordinary differential equations

$$\frac{d^2 f(x)}{dx^2} - \lambda f(x) = 0$$

$$\frac{d^2 g(y)}{dy^2} + \lambda g(y) = 0$$

These are easily solved. The nature of the solution depends on the sign of λ, which in turn depends on the *boundary conditions* satisfied by z. This aspect is beyond the scope of this book, but we have already seen enough to appreciate that we have here a valuable method of solving PDEs.

EXERCISE ON 4.4

Apply the method of separation of variables to the following PDEs:

(i) $\dfrac{\partial^2 z}{\partial x^2} - \dfrac{\partial^2 z}{\partial y^2} = 0$

(ii) $\dfrac{\partial^2 z}{\partial x^2} = \dfrac{\partial z}{\partial y}$

4.5 The chain rules

As mentioned in Section 4.3, we need a way of dealing with the situation where both x and y vary simultaneously in the function $f(x, y)$. One approach is to regard x and y as functions of an independent parameter t, $x = x(t)$, $y = y(t)$. Then $z = f(x(t), y(t)) = z(t)$ also becomes a function of t, which we can differentiate in the normal way, with respect to t, to give dz/dt. But there is still the question of how this is related to the partial derivatives f_x, f_y. There is also the question of the conditions under which $z = f(x, y)$ can be differentiated in this way. It might seem reasonable to expect that for this to be the case f_x, f_y must both exist – but in fact it turns out that *this is not enough*. We must also require that they are both *continuous*. This is a subtlety about differentiation of functions of more than one variable which requires a generalization of what we mean by 'differentiability' of such functions. Here we will give a brief outline of what is involved, leaving detailed discussion and proof to Chapter 5.

If we look at the case of a function of a single variable, $f(x)$, then this is 'differentiable' if the limit

$$\lim_{\Delta x \to 0} \left(\frac{f(x + \Delta x) - f(x)}{\Delta x} \right) = f'(x)$$

exists. This is not easily generalized to functions of two or more variables, because it is not clear what would replace Δx in the denominator. We therefore rephrase the above definition of differentiability in the following form:

● *Differentiability: functions of one variable* ─────────

f is *differentiable* at x if there is a number $f'(x)$ and a function $\varepsilon(\Delta x)$ such that

$$f(x + \Delta x) - f(x) = f'(x)\Delta x + \varepsilon(\Delta x)\Delta x$$

where

$$\lim_{\Delta x \to 0} \varepsilon(\Delta x) = 0$$

This can be generalized to functions of more than one variable, leading to the definition:

● *Differentiability: functions of more than one variable* ─────────

$z = f(x, y)$ is *differentiable* at (x, y) if $f_x(x, y), f_y(x, y)$ both exist in a neighbourhood of (x, y) and there exist functions $\varepsilon_1(\Delta x, \Delta y)$, $\varepsilon_2(\Delta x, \Delta y)$ such that

$$\Delta f = f(x + \Delta x, y + \Delta y) - f(x, y)$$

$$= f_x(x, y)\, \Delta x + f_y(x, y)\Delta y + \varepsilon_1(\Delta x, \Delta y)\Delta x + \varepsilon_2(\Delta x, \Delta y)\Delta y$$

where

$$\lim_{(\Delta x, \Delta y) \to (0,\, 0)} \varepsilon_1(\Delta x, \Delta y) = \lim_{(\Delta x, \Delta y) \to (0,\, 0)} \varepsilon_2(\Delta x, \Delta y) = 0$$

Essentially, this amounts to a linear approximation to Δf in terms of the first partial derivatives. And it only uses terms with which we are already familiar. Note that the notion of x and y varying *together* has now quietly slipped in by the use of the limit $(\Delta x, \Delta y) \to (0, 0)$, whereas in the definition of the partial derivatives we used $\Delta x \to 0$ or $\Delta y \to 0$ *separately*. We will have more to say on this in Chapter 5, where it is proved that $z = f(x, y)$ is differentiable in the above sense if f, f_x, f_y are all defined and continuous in a neighbourhood of (x, y). For the moment just assume this to be the case. Then dz/dt is defined, and, as we will show in Chapter 5, is given by a 'chain rule' analogous to the 'function of a function' rule of elementary calculus:

$$\frac{dy}{dx} = \frac{dy}{du}\frac{du}{dx}$$

Note that this simple expression already signals a common abuse of notation which we will continue to apply throughout the calculus of several variables. On the left-hand side y is regarded as a function of x and on the right-hand side it represents a function of u obtained by replacing x in terms of u. Strictly speaking, one should write

$$\frac{dy(x)}{dx} = \frac{dy(x(u))}{du}\frac{du(x)}{dx}$$

but this is clumsy and so we use the notational licence of denoting $y(x)$ and $y(x(u))$ by y. We do a similar thing in partial differentiation, but more care is needed because of the additional variables.

In the case of a function of two variables there are two forms that the chain rule can take.

● Chain rule 1

Let $z = f(x, y)$ be differentiable and suppose that $x = x(t)$ and $y = y(t)$. Assume that dx/dt and dy/dt exist and are continuous. Then we can also write z as a function of t, $z = z(x(t), y(t)) \equiv z(t)$, and its derivative with respect to t is given by

$$\frac{dz}{dt} = \frac{\partial z}{\partial x}\frac{dx}{dt} + \frac{\partial z}{\partial y}\frac{dy}{dt} = f_x\frac{dx}{dt} + f_y\frac{dy}{dt}$$

● Chain rule 2

Let $z = f(x, y)$ be differentiable and suppose that x and y are functions of the two variables r and s, $x = x(r, s)$, $y = y(r, s)$. Assume that $\partial x/\partial r$, $\partial x/\partial s$, $\partial y/\partial r$ and $\partial y/\partial s$ all exist and are continuous. Then z can be written as a function of r and s, and

$$\frac{\partial z}{\partial r} = \frac{\partial z}{\partial x}\frac{\partial x}{\partial r} + \frac{\partial z}{\partial y}\frac{\partial y}{\partial r}$$

$$\frac{\partial z}{\partial s} = \frac{\partial z}{\partial x}\frac{\partial x}{\partial s} + \frac{\partial z}{\partial y}\frac{\partial y}{\partial s}$$

Notice again the abuse of notation referred to above. The *z*s on either side of the equations given represent the same quantity *z* expressed in terms of the appropriate variables.

Chain rule 1 clearly follows from chain rule 2 by putting *s* = constant, and so we only need to prove the latter. Also, we only need to derive the first equation, as the second follows identically. We give the proof in Chapter 5, as it uses the discussion of differentiability included there. Here we are just going to use the technology that the chain rules provide.

PROBLEM 6

Let $z = f(x, y) = \frac{1}{3}\pi x^2 y$ and let $x = t$, and $y = e^t$. Calculate dz/dt, and obtain the rate of increase of *z* with respect to *t* at *t* = 1.

We have

$$\frac{dz}{dt} = \frac{\partial z}{\partial x}\frac{dx}{dt} + \frac{\partial z}{\partial y}\frac{dy}{dt} = \left(\frac{2}{3}\pi xy\right)(1) + \left(\frac{1}{3}\pi x^2\right)(e^t)$$

$$= \frac{2}{3}\pi te^t + \frac{1}{3}\pi t^2 e^t = \frac{\pi}{3}te^t(2 + t)$$

At *t* = 1 we have

$$\frac{dz}{dt} = \pi e$$

PROBLEM 7

Let $z = \cos(xy)$ and suppose $x = rs$, $y = r + s$. Calculate $\partial z/\partial r$ and $\partial z/\partial s$ – do you see any short-cuts?

$$\frac{\partial z}{\partial r} = \frac{\partial z}{\partial x}\frac{\partial x}{\partial r} + \frac{\partial z}{\partial y}\frac{\partial y}{\partial r} = (-y\sin(xy))(s) + (-x\sin(xy))(1)$$

$$= -(r + s)s\sin(rs(r + s)) - rs\sin(rs(r + s))$$

$$= -s(2r + s)\sin(rs(r + s))$$

$$\frac{\partial z}{\partial s} = \frac{\partial z}{\partial x}\frac{\partial x}{\partial s} + \frac{\partial z}{\partial y}\frac{\partial y}{\partial s} = -r(2s + r)\sin(rs(r + s))$$

Note that $\partial z/\partial s$ can be obtained from $\partial z/\partial r$ simply by swapping round *s* and *r*, since both *x* and *y* are symmetric in *s* and *r*.

In practical applications of the chain rules, the calculations can become quite complicated, particularly when applying the results to second-order derivatives. In such cases it is often possible to take advantage of *partial differential operator* techniques. For example, in using the chain rule to transform from one set of variables to another it may be possible to define transformed differential operators which simplify calculations. An example will illustrate the approach.

PROBLEM 8

Let f be a function of x, y and suppose that (x, y) are transformed to new variables (r, θ) by

$$x = \sqrt{r}\cos(\theta/2) \quad y = \sqrt{r}\sin(\theta/2)$$

Show that

(i) $2r\dfrac{\partial f}{\partial r} = x\dfrac{\partial f}{\partial x} + y\dfrac{\partial f}{\partial y}$

(ii) $2\dfrac{\partial f}{\partial \theta} = x\dfrac{\partial f}{\partial y} - y\dfrac{\partial f}{\partial x}$

Treating (i) and (ii) as derivative operator relations, show that

(iii) $4\left(r\dfrac{\partial^2 f}{\partial r^2} + \dfrac{\partial f}{\partial r} + \dfrac{1}{r}\dfrac{\partial^2 f}{\partial \theta^2}\right) = \dfrac{\partial^2 f}{\partial x^2} + \dfrac{\partial^2 f}{\partial y^2}$

(i) We have, by the chain rule,

$$\frac{\partial f}{\partial r} = \frac{\partial f}{\partial x}\frac{\partial x}{\partial r} + \frac{\partial f}{\partial y}\frac{\partial y}{\partial r} = \frac{1}{2\sqrt{r}}\cos\left(\frac{\theta}{2}\right)\frac{\partial f}{\partial x} + \frac{1}{2\sqrt{r}}\sin\left(\frac{\theta}{2}\right)\frac{\partial f}{\partial y}$$

or

$$2r\frac{\partial f}{\partial r} = \sqrt{r}\cos\left(\frac{\theta}{2}\right)\frac{\partial f}{\partial x} + \sqrt{r}\sin\left(\frac{\theta}{2}\right)\frac{\partial f}{\partial y}$$

$$= x\frac{\partial f}{\partial x} + y\frac{\partial f}{\partial y}$$

(ii) Similarly, you can confirm that

$$2\frac{\partial f}{\partial \theta} = x\frac{\partial f}{\partial y} - y\frac{\partial f}{\partial x}$$

Now the thing to notice here is that the results (i) and (ii) are true for any differentiable function f. We may therefore express them formally as *operator identities*:

$$2r\frac{\partial}{\partial r} \equiv x\frac{\partial}{\partial x} + y\frac{\partial}{\partial y}$$

$$2\frac{\partial}{\partial \theta} \equiv x\frac{\partial}{\partial y} - y\frac{\partial}{\partial x}$$

it being understood that these have a meaning when applied to appropriate functions and their transforms. The power of such operators is illustrated in the final part of the problem.

(iii) Notice that

$$4\left(r\frac{\partial^2 f}{\partial r^2} + \frac{\partial f}{\partial r} + \frac{1}{r}\frac{\partial^2 f}{\partial\theta^2}\right) = \frac{1}{r}\left(2r\frac{\partial}{\partial r}\left(2r\frac{\partial f}{\partial r}\right) + 2\frac{\partial}{\partial\theta}\left(2\frac{\partial f}{\partial\theta}\right)\right)$$

or in terms of the operators defined above,

$$= \frac{1}{r}\left(2r\frac{\partial}{\partial r}\left(2r\frac{\partial}{\partial r}\right) + 2\frac{\partial}{\partial\theta}\left(2\frac{\partial}{\partial\theta}\right)\right)f$$

Substituting for the operators gives

$$\frac{1}{r}\left(\left(x\frac{\partial}{\partial x} + y\frac{\partial}{\partial y}\right)\left(x\frac{\partial}{\partial x} + y\frac{\partial}{\partial y}\right) + \left(x\frac{\partial}{\partial y} - y\frac{\partial}{\partial x}\right)\left(x\frac{\partial}{\partial y} - y\frac{\partial}{\partial x}\right)\right)f$$

$$= \frac{1}{r}\left((x^2 + y^2)\frac{\partial^2}{\partial x^2} + (x^2 + y^2)\frac{\partial^2}{\partial y^2}\right)$$

on cancelling like terms,

$$= \frac{\partial^2 f}{\partial x^2} + \frac{\partial^2 f}{\partial y^2}$$

on noticing that $r = x^2 + y^2$.

The chain rules may be extended in an obvious way to functions of more variables, expressed in terms of more parameters – see Exercises 4.5 for examples.

Although you might not realize it, it is possible that you have already made some use of the two-variable chain rule in ordinary differentiation – namely when you did *implicit differentiation*. Thus, suppose the equation $F(x, y) = 0$ defines y implicitly as a differentiable function of x, say $y = f(x)$, and suppose that $F(x, y)$ is differentiable. Then since

$$z = F(x, y) = 0$$

we have

$$\frac{dz}{dx} = 0 = F_x\frac{dx}{dx} + F_y\frac{dy}{dx}$$

$$= F_x + F_y\frac{dy}{dx}$$

and so

$$\frac{dy}{dx} = -\frac{F_x}{F_y}$$

PROBLEM 9

If $x^2 + 2xy + y^2 = 0$, obtain dy/dx using the chain rule – and confirm alternatively.

We have, with $z = x^2 + 2xy + y^2 = 0$,

$$\frac{dz}{dx} = (2x + 2y) + (2x + 2y)\frac{dy}{dx} = 0$$

so

$$\frac{dy}{dx} = -\frac{2x + 2y}{2x + 2y} = -1$$

Or, note that the equation implies $x + y = 0$, etc.

By the way, while we are on the subject of implicit functions, did it occur to you to question whether we *can* always solve an 'implicit' equation such as $F(x, y) = 0$ to obtain $y = f(x)$ explicitly? It is easy to see that we cannot always do this *in practice* – for example,

$$ye^x + xe^y = 0$$

cannot be solved explicitly for x or y. But, is it possible *in principle*? And what can we say about the properties of such a solution? These questions are answered by one of the key theorems of calculus:

● Implicit function theorem ————————————————

If $F(x, y)$ has continuous partial derivatives in the neighbourhood of some point (x_0, y_0) at which $F_x(x_0, y_0) = 0$ and $F_y(x_0, y_0) \neq 0$, then there is a function $f(x)$, and an interval about the point x_0 in which

$$F(x, f(x)) = 0$$

and $f'(x)$ exists and is continuous.

The proof of this result is beyond the scope of this book, but the theorem is fundamental to advanced calculus. (See Pedrick, 1994.)

EXERCISES ON 4.5

1. Let $w = f(x, y, z)$ and let (x, y, z) be points on a surface parametrized by (r, s), so that $x = x(r, s)$, $y = y(r, s)$, $z = z(r, s)$. Write down the chain rule for the partial derivatives of w with respect to r and s.
2. Given $z = e^{x+y} \cos(xy)$ and $x = t$, $y = t^2$, evaluate dz/dt using the chain rule, and also by expressing z as a function of t and differentiating with respect to t. Compare your results.
3. Given $z = x^2y$ and $x = s \cos r$, $y = r \sin s$, evaluate $\partial z/\partial r$ and $\partial z/\partial s$.
4. If $x^3 + 2xy \sin(x + y) = 0$, obtain dy/dx as a function of x and y.

4.6 The total differential

If, in chain rule 1, we 'cancel' the dt throughout we get, for $z = f(x, y)$,

$$dz = f_x \, dx + f_y \, dy$$

This is a relation between 'differentials' dx, dy, dz, and dz is called the *total differential* of z. This is a *definition, not an approximation.* It is important *symbolically*, not *numerically*. However, there is an approximate analogue to it which arises simply from the definition of differentiability. We will give more details in Chapter 5, but for now simply note that if $z = f(x, y)$ is differentiable then we can write for the *increment* $\Delta f = f(x + \Delta x, y + \Delta y) - f(x, y)$

$$\Delta f = f_x \, \Delta x + f_y \, \Delta y + \varepsilon_1(\Delta x, \Delta y)\Delta x + \varepsilon_2(\Delta x, \Delta y)\Delta y$$

So, if Δx, Δy are small we can write

$$\Delta f \approx f_x \Delta x + f_y \Delta y$$

This form alone motivates the definition of the total differential, df, but the increment approximation form stands in its own right as a useful approximation to the change in f due to changes Δx, Δy in x and y respectively. We can in fact get an explicit idea of what is going on by looking at a few examples.

PROBLEM 10

Obtain expressions for Δz in the cases

(i) $z = x + y$ (ii) $z = xy$
(iii) $z = x^2 + y^2$ (iv) $z = x \cos y$

What happens if Δx, Δy are so small that products of them can be neglected?

(i) This is not so bad:

$$\Delta z = x + \Delta x + y + \Delta y - (x + y) = \Delta x + \Delta y$$

This result is exact. There are no products of Δx, Δy.

(ii) This is a little more complex:

$$\Delta z = (x + \Delta x)(y + \Delta y) - xy$$
$$= y\Delta x + x\Delta y + \Delta x \Delta y$$

This is exact. However, if we neglect $\Delta x \Delta y$ then we get an *approximation*:

$$\Delta z \approx y\Delta x + x\Delta y$$

Notice that since $f_x = y$ and $f_y = x$ this may be written

$$\Delta z \approx f_x \Delta x + f_y \Delta y$$

(iii) $\Delta z = (x + \Delta x)^2 + (y + \Delta y)^2 - x^2 - y^2$
         $= 2x\Delta x + 2y\Delta y + (\Delta x)^2 + (\Delta y)^2$
         $\approx 2x\Delta x + 2y\Delta y$

if we neglect the Δ-products. Again, notice that this is

$$\Delta z \approx f_x \Delta x + f_y \Delta y$$

(iv) $\Delta z = (x + \Delta x) \cos(y + \Delta y) - x \cos y$

Not so easy! Suppose we push it a bit:

$$\Delta z = (x + \Delta x)(\cos y \cos \Delta y - \sin y \sin \Delta y) - x \cos y$$

Now since Δ-products may be ignored, we can use the approximations

$$\cos \Delta y \approx 1 \qquad \sin \Delta y \approx \Delta y$$

to get

$$\Delta z \approx (x + \Delta x)(\cos y - \Delta y \sin y) - x \cos y$$
$$\approx (\cos y) \Delta x - (x \sin y) \Delta y$$

on again neglecting Δ-products.

I am sure you have already noticed that this again takes the form $f_x \Delta x + f_y \Delta y$.

These examples illustrate the result

$$\Delta z \simeq f_x \Delta x + f_y \Delta y$$

which we obtained from the definition of differentiability.

The above discussion is not rigorous, but sufficiently illustrative for us to defer justification until Section 5.2. For now, we accept that if Δx, Δy are small numbers and $z = f(x, y)$ is differentiable then

$$\Delta z \simeq f_x \Delta x + f_y \Delta y \tag{4.1}$$

As stated earlier this is an *approximation* which generally improves as Δx, Δy get smaller and smaller. It is the two-dimensional generalization of the single-variable approximation

$$\Delta y \simeq y' \Delta x$$

It leads naturally to the definition of the total differential of $f(x, y)$ given earlier,

$$df = f_x \, dx + f_y \, dy$$

I repeat that this *is* a definition, and *not* an approximation – there is no need for dx, dy to be 'small'. Also, note that while Δx, Δy, Δz are *numbers*, dx, dy, df are *not* – they are *differentials*. This is not nit-picking, it is an extremely important distinction. Some textbooks ask us to believe that if Δx, Δy are small, then

$$\Delta z \simeq dz$$

This, however, is like equating apples and pears. What is meant is that the formal *expression* for Δz as a function of Δx, Δy is approximately equal to that for dz if Δx, Δy are small enough.

The generalization of the total differential to three or more variables is obvious:

$$dz = \sum_{i=1}^{n} \frac{\partial z}{\partial x_i} \, dx_i$$

If you have any curiosity at all you will be feeling distinctly uneasy at this point. If these dx, dy, dz are *not* small numerical quantities, then *what are they* – and what *is* their relationship to Δx, Δy, Δz, or to the dx, dy in dy/dx, for example? I have called them differentials – but what does that mean? In fact, the theory of such differentials is subtle, advanced, and until relatively recently contentious.

Things like dx, dy are often described as *infinitesimals* – they are not numbers, but are regarded as variables which can be 'as small as you wish' but never actually zero. In calculations they can never be zero, because you might want to divide by them, but eventually you are going to put them to zero. It has long been the tale of 'standard' analysis that all this is made perfectly respectable by the use of limits, and that there is no need for such shenanigans with infinitesimals. This is despite the fact that applied mathematicians wield them with great power all the time. However, during the 1960s mathematicians devised *non-standard analysis* in which actual infinitesimals do exist – as constants, not variables. In non-standard analysis the equivalent of dx, dy are treated as 'hyperreal' numbers which can be manipulated analogously to ordinary numbers. Such analysis has found applications not only in mathematics but also in applied mathematics and engineering. For a very readable account of the subject, *see* Stewart (1996). In this book we are going to stick firmly to standard analysis and work with infinitesimals or differentials dx, dy and increments Δx, Δy.

The total differential raises an interesting point concerning differentials in general, which provides a bridge to many other mathematical topics – differential equations, integration and mechanics. It is something to which we will return in Chapter 10, but we can raise it here.

Whatever they are, dx, dy may be combined linearly in expressions such as $P(x, y)dx + Q(x, y)dy$, where P and Q are appropriately well-behaved functions. An expression of this type is called a *differential form*. The total differential is itself a differential form:

$$df = f_x \, dx + f_y \, dy$$

It is special in that the linear combination can itself be expressed as another differential, df. The obvious question arises about *how* special this is – under what conditions can an expression $Pdx + Qdy$ be written as a total differential of some function f? When this is possible we say that $Pdx + Qdy$ is an *exact differential*. The necessary and sufficient condition for

$$P(x, y) \, dx + Q(x, y) \, dy$$

to be an exact differential is that P, Q are continuous functions with continuous partial derivatives satisfying

$$\frac{\partial P}{\partial y} = \frac{\partial Q}{\partial x}$$

A proof is given in Cox (1996). It is a constructive proof in that it enables us actually to determine f. The proof also reveals that the essential underlying principle involved is simply the equality of mixed derivatives $f_{xy} = f_{yx}$. This is often the case in advanced mathematics – apparently sophisticated and technical results sometimes have mundane, but fundamental, roots. Another example of this is the underlying role of the *fundamental theorem of integral calculus* in the development of Green's, Stokes's and the divergence theorems, which is where we will return to the topic of the total differential. Expressing, where possible, $Pdx + Qdy$ in the form df enables us to take advantage of the fundamental theorem of integral calculus and integrate directly.

PROBLEM 11

Find the total differential dz when $z = \ln(\sin(xy))$.

$$\frac{\partial z}{\partial x} = \frac{y \cos(xy)}{\sin(xy)} = y \cot(xy)$$

$$\frac{\partial z}{\partial y} = x \cot(xy) \quad \text{by symmetry}$$

So

$$dz = \frac{\partial z}{\partial x}dx + \frac{\partial z}{\partial y}dy = \cot(xy)(y\,dx + x\,dy)$$

PROBLEM 12

Find the percentage error in $z = x^\alpha y^\beta$ in terms of the percentage errors in x and y.

First the long way. Using the approximation for increments,

$$\Delta z \simeq \alpha x^{\alpha-1}y^\beta \Delta x + \beta x^\alpha y^{\beta-1}\Delta y$$

So the relative error in z is

$$\frac{\Delta z}{z} = \frac{\alpha x^{\alpha-1}y^\beta}{z}\Delta x + \frac{\beta x^\alpha y^{\beta-1}}{z}\Delta y$$

$$= \alpha\frac{x^{\alpha-1}y^\beta}{x^\alpha y^\beta}\Delta x + \beta\frac{x^\alpha y^{\beta-1}}{x^\alpha y^\beta}\Delta y$$

$$= \alpha\frac{\Delta x}{x} + \beta\frac{\Delta y}{y}$$

So the percentage error is

$$\frac{\Delta z}{z} \times 100 = \left(\alpha\frac{\Delta x}{x} + \beta\frac{\Delta y}{y}\right)100$$

$$= \alpha\left(\frac{\Delta x}{x} \times 100\right) + \beta\left(\frac{\Delta y}{y} \times 100\right)$$

$$= \alpha\text{ percentage error in }x + \beta\text{ percentage error in }y$$

Notice the pattern.
   Now the easy way. Take logs:

$$\ln z = \ln(x^\alpha y^\beta) = \alpha \ \ln x + \beta \ln y$$

Now take the total differential,

$$d(\ln z) = \frac{dz}{z} = \alpha\frac{dx}{x} + \beta\frac{dy}{y}$$

and proceed as before.

EXERCISE ON 4.6

If $r = \sqrt{x^2 + y^2 + z^2}$, obtain an expression for the change, Δr, in r due to small changes Δx, Δy, Δz in x, y, z respectively. Hence find an approximate value for $\sqrt{(4.02)^2 + (3.99)^2 + (1.99)^2}$.

FURTHER EXERCISES

1. Find the first- and second-order partial derivatives of the following functions:

   (i) $x + 3y$ (ii) $2x^2 + 4y^2$ (iii) $3x^4 - 4x^2y^2 + 5y^3$

   (iv) xye^{x+y} (v) $\cos(x + y^2)$ (vi) $\sin(x)\cos(xy)$

   (vii) $\dfrac{x-y}{x+y}$ (viii) $\ln\left(\dfrac{x^2+y}{2x+3y}\right)$ (ix) $\dfrac{e^{x+y}}{x+y}$

   (x) $e^{x+y}\cos(x^2 + y)$

2. Find the first- and second-order partial derivatives of the following functions:

   (i) $x + y^2 + z^3$ (ii) $\dfrac{xyz}{x+y+z}$ (iii) $e^{x+y^2+z^3}$

   (iv) $\ln\left(\dfrac{x+y+2z}{x^2+y^2}\right)$ (v) $\dfrac{x}{y}+\dfrac{y}{z}+\dfrac{z}{x}$

3. Verify the following solutions of the Laplace equation,

$$\frac{\partial^2 f}{\partial x^2} + \frac{\partial^2 f}{\partial y^2} = 0$$

   (i) $f(x, y) = x^2 - y^2$ (ii) $f(x, y) = e^{-2y}\cos 2x$

   (iii) $f(x, y) = e^x(x \cos y - y \sin y)$ (iv) $f(x, y) = x^3 - 3xy^2$

4. Show that $T(x, t) = ae^{-b^2 t}\cos bx$, where a and b are arbitrary constants, satisfies the *heat equation*

$$\frac{\partial T}{\partial t} = \frac{\partial^2 T}{\partial x^2}$$

5. If $z = x^3 \sin y$, verify that

$$\frac{\partial^2 z}{\partial x \partial y} = \frac{\partial^2 z}{\partial y \partial x}$$

$$\left(\frac{\partial z}{\partial x}\right)_y \left(\frac{\partial x}{\partial y}\right)_z \left(\frac{\partial y}{\partial z}\right)_x = -1$$

$$\frac{\partial^{10} z}{\partial x^4 \partial y^6} = \frac{\partial^{10} z}{\partial y^6 \partial x^4} = 0$$

6. If $f(x, y, z) = x^2 e^{yz}\cos(y^2 z)$, find

$$\frac{\partial^{20} f}{\partial x^3 \partial y^7 \partial z^{10}}$$

7. Find the values of n so that $f(r, \theta) = r^n(3 \cos^2\theta - 1)$ satisfies the equation

$$\frac{\partial}{\partial r}\left(r^2\frac{\partial f}{\partial r}\right) + \frac{1}{\sin\theta}\frac{\partial}{\partial\theta}\left(\sin\theta\frac{\partial f}{\partial\theta}\right) = 0 \qquad (r \neq 0)$$

8. Show that $T = e^{-x}\sin(x - 2t)$ satisfies the heat equation

$$\frac{\partial T}{\partial t} = \frac{\partial^2 T}{\partial x^2}$$

9. If $\phi = \cos(x + y) + \sin(x - y)$, show that

$$\frac{\partial^2\phi}{\partial x^2} - \frac{\partial^2\phi}{\partial y^2} = 0$$

This is the *wave equation*.

10. Show that if $w = \tan^{-1}(x/y)$ then

$$\frac{\partial^2 w}{\partial x^2} + \frac{\partial^2 w}{\partial y^2} = 0$$

11. (i) If $z = 4x^2 + xy + y^2$ and x and y vary with time t according to $x = 1 + \cos t$ and $y = 2 \sin t$, evaluate dz/dt directly and by using the chain rule.

    (ii) If $z = xy^2 + xy - y^4$ and $x = 2t$, $y = t^3$, then evaluate dz/dt directly and by the chain rule.

    (iii) The surface area S of the curved surface of a cone of base radius r and height h is given by $S = \pi r^2 + \pi r \sqrt{r^2 + h^2}$. Find the rate at which S is increasing when $h = 3$ cm and $r = 4$ cm if h and r are both increasing at 0.3 cm s^{-1}.

12. (i) From the ideal gas law $PV = nRT$, where nR are constants, use the total differential to estimate the percentage change in pressure, P, if the temperature, T, is increased by 3% and the volume V is increased by 5%.

    (ii) The length and width of a rectangle are measured with maximum errors of $\varepsilon\%$. Estimate the maximum percentage error in

      (a) the area of the rectangle;

      (b) the diagonal of the rectangle.

    (iii) The total resistance of three resistors R_1, R_2, R_3 in parallel is given by R, where

$$\frac{1}{R} = \frac{1}{R_1} + \frac{1}{R_2} + \frac{1}{R_3}$$

If R_1, R_2, R_3 are measured, in consistent units, as 6, 8, 12 respectively, with respective maximum tolerances of ±0.1, ±0.03, ±0.15, estimate the maximum possible error in R.

13. If ϕ is a function of u and v and

$$u = x^3 - 3xy^2$$
$$v = 3x^2 y - y^3$$

show that

$$x\frac{\partial \phi}{\partial x} + y\frac{\partial \phi}{\partial y} = 3\left(u\frac{\partial \phi}{\partial u} + v\frac{\partial \phi}{\partial v}\right)$$

14. Show that if V is a function of x, y and

$$x = e^u \cos \theta$$
$$y = e^u \sin \theta$$

then

$$\frac{\partial^2 V}{\partial u^2} + \frac{\partial^2 V}{\partial \theta^2} = e^{2u}\left(\frac{\partial^2 V}{\partial x^2} + \frac{\partial^2 V}{\partial y^2}\right)$$

15. Let r and θ be standard polar coordinates, i.e. $x = r \cos \theta$, $y = r \sin\theta$. Show that

$$(\partial x/\partial r)_\theta = \cos \theta \quad (\partial r/\partial x)_y = \cos \theta$$
$$(\partial x/\partial \theta)_r = r \sin \theta \quad (\partial \theta/\partial x)_y = -r^{-1} \sin \theta$$

Note that

$$(\partial x/\partial r)_\theta \neq [(\partial r/\partial x)_y]^{-1}$$

$$(\partial x/\partial \theta)_r \neq [(\partial \theta/\partial x)_y]^{-1}$$

However, by expressing x as a function of r and y only, deduce that

$$(\partial x/\partial r)_y = r(r^2 - y^2)^{-1/2} = 1/\cos\theta$$

Hence

$$(\partial x/\partial r)_y = [(\partial r/\partial x)_y]^{-1}$$

Similarly, show that

$$(\partial x/\partial \theta)_y = [(\partial \theta/\partial x)_y]^{-1}$$

16. If $x = e^u + v$, $y = e^{-u} + v$ and $f(x, y) = xy$, use the chain rule for partial derivatives to show that

$$\frac{\partial f}{\partial u} = 2v \sinh u$$

$$\frac{\partial f}{\partial v} = 2 \cosh u + 2v$$

Express f in terms of u and v and verify these formulae directly.

17. If $x = u^2 + v^2$ and $y = u/v$ and f is an arbitrary function of x and y, calculate $\partial f/\partial u$ and $\partial f/\partial v$. Solve these expressions to find $\partial f/\partial x$ and $\partial f/\partial y$ in terms of u, v, $\partial f/\partial u$ and $\partial f/\partial v$. Show that

$$4\frac{\partial^2 f}{\partial x^2} = (u^2 + v^2)^{-2}\left\{u^2\frac{\partial^2 f}{\partial u^2} + 2uv\frac{\partial^2 f}{\partial u \partial v} + v^2\frac{\partial^2 f}{\partial v^2} - u\frac{\partial f}{\partial u} - v\frac{\partial f}{\partial v}\right\}$$

18. The variables x and y are related to new variables X and Y by the equations

$$X = xy \qquad Y = 1/y$$

Show that

(i) $y\dfrac{\partial}{\partial y} \equiv X\dfrac{\partial}{\partial X} - Y\dfrac{\partial}{\partial Y}$

(ii) $x\dfrac{\partial}{\partial x} - y\dfrac{\partial}{\partial y} \equiv Y\dfrac{\partial}{\partial Y}$

(iii) $y\dfrac{\partial f}{\partial y}\left(x\dfrac{\partial f}{\partial x} - y\dfrac{\partial f}{\partial y}\right) = Y\dfrac{\partial f}{\partial Y}\left(X\dfrac{\partial f}{\partial X} - Y\dfrac{\partial f}{\partial Y}\right)$

(iv) $y\left(x\dfrac{\partial^2 f}{\partial x \partial y} - y\dfrac{\partial^2 f}{\partial y^2} - \dfrac{\partial f}{\partial y}\right) = Y\left(X\dfrac{\partial^2 f}{\partial x \partial y} - Y\dfrac{\partial^2 f}{\partial Y^2} - \dfrac{\partial f}{\partial Y}\right)$

19. Prove that differentiable functions of several variables are most sensitive to small changes in the variable which generates the largest partial derivative. Illustrate this in the case of changes in the volume of a right circular cylinder.

20. Essay or discussion topics:

   (i) The key principle in the use of separation of variables to solve PDEs.

   (ii) Extending the chain rules to functions of more than two variables.

   (iii) The difference between Δx and $\mathrm{d}x$.

   (iv) The total differential, implicit differentiation, approximations and exact differentials.

5 • Differentiability: Analytical Aspects

5.1 Introduction

In Chapter 4 we started to look at the calculus of two or more variables proper – situations where all variables may change at once. The point was made that the theory of such things is slightly more subtle than the calculus of functions of a single variable. Essentially, this is because it is not clear what should replace Δx in the definition

$$f'(x) = \lim_{\Delta x \to 0} \left[\frac{f(x + \Delta x) - f(x)}{\Delta x} \right]$$

In this case Δx is a single real variable, which, because it is under the limit, is never actually going to reach the value of zero and so division by it is always defined. In the case of a function of two variables we might envisage a generalization to something like

$$\lim_{(\Delta x,\, \Delta y) \to (0,0)} \left[\frac{f(x + \Delta x,\, y + \Delta y) - f(x,\, y)}{?} \right]$$

but what would be the generalization of Δx in the denominator? It could not be $(\Delta x,\, \Delta y)$ – you cannot divide by an ordered pair, just as you cannot divide by a vector. As noted in Section 4.5, what we do is to modify the single-variable definition slightly. This is a standard trick in mathematics when we need to generalize a result to new circumstances such as higher dimensions – we rearrange the statement of the result to a form which is most easily generalized. In this particular case, the rearrangement and its generalization are in fact surprisingly rich in new suggestive ideas, which we can explore by digging a bit deeper into the above definition.

But first, some work.

PROBLEM 1

Write down all the interpretations that you can think of for the derivative of a function of a single variable.

- $f'(x)$ is the limit defined earlier.
- $f'(x)$ is the rate of change of $f(x)$ as x increases.
- $f'(x)$ is the gradient of the curve $y = f(x)$ at x, defined as the slope of the tangent to the curve $y = f(x)$ at x.
- If Δx is very small, then

$$f(x + \Delta x) \simeq f(x) + f'(x)\, \Delta x$$

You may have more, but these will do. Essentially they can be summarized succinctly as

- derivative as a *limit*;
- derivative as a 'velocity';
- derivative as the gradient of a *curve*;
- derivative as providing a *linear* (*in* Δx) *approximation* for a function.

All these viewpoints are clearly related. The limit is the mathematical means by which each of the others is defined. The gradient is essentially the geometric manifestation of velocity. The linear approximation amounts essentially to replacing the curve $f(x)$ at a point by a short segment of the tangent to the curve at that point. In this chapter we will focus on the purely mathematical and geometrical aspects, relating the underlying concept of differentiability to limits, tangent gradients and linear approximations. For each of these is easily generalized to functions of two (or more variables) – the limit idea we extended in Chapter 3, the tangent line extends naturally to tangent plane, and the linear approximation of a curve by a straight line extends naturally to the 'linear approximation' of a surface by a portion of the tangent plane. All of these viewpoints provide useful tools for studying the calculus of two or more variables.

First, we will look more carefully at the new definition of differentiability.

5.2 Differentiability: a definition

First, let us see why the existence of partial derivatives is not really enough for true differentiability of functions of more than one variable. Remember that in the case of a function of a single variable, *differentiability implies continuity*. That is, if a function $f(x)$ is differentiable, then it is also continuous. Graphically this is obviously necessary, since a curve with a break in it cannot have a tangent at the break. The question is, can we extend this idea to two variables? If we regard a function $f(x, y)$ as 'differentiable' if its partial derivatives f_x, f_y exist then the answer is emphatically *no*. That is, it is possible for $f(x,y)$ to have partial derivatives at a point at which it is not continuous.

PROBLEM 2

Show that f_x, f_y both exist at the origin for the function

$$f(x, y) = \begin{cases} \dfrac{xy}{x^2 + y^2} & (x, y) \neq (0, 0) \\ 0 & (x, y) = (0, 0) \end{cases}$$

then compare with Problem 9 of Chapter 3.

At the origin we have

$$f_x(0, 0) = \lim_{h \to 0} \left[\frac{f(0 + h, 0) - f(0, 0)}{h} \right]$$

$$= \lim_{h \to 0} \left[\frac{(0+h) \cdot 0}{h^2 + 0^2} \middle/ h \right] = \lim_{h \to 0} \frac{0}{h} = 0$$

Similarly we find $f_y(0, 0) = 0$. Thus the partial derivatives exist – but we know from Problem 9 of Chapter 3 that $f(x, y)$ is not continuous at $(0, 0)$.

So, the existence of the partial derivatives alone does not fulfil what we would expect of differentiability. The reason for this is actually quite obvious. The existence of f_x, f_y depends only on points of the form $(x_0 + h, y_0)$, $(x_0, y_0 + k)$, respectively, whereas continuity depends on points in all directions $(x_0 + h, y_0 + k)$ (a similar situation arises in the case of continuity – continuity in each variable separately does not guarantee continuity of the function; *see* Section 3.3). We must therefore find a more complete definition of differentiability. As mentioned in Section 4.5, we do this by taking the lead from the alternative definition of the differentiability of a function of a single variable:

● *Differentiability: functions of one variable* ─────────

f is *differentiable* at x if there is a number $f'(x)$ and a function $\varepsilon(\Delta x)$ such that

$$f(x + \Delta x) - f(x) = f'(x)\Delta x + \varepsilon(\Delta x)\Delta x \qquad (5.1)$$

where

$$\lim_{\Delta x \to 0} \varepsilon(\Delta x) = 0$$

This definition is an alternative expression for the existence of the derivative. It illustrates very effectively the role of the derivative $f'(x)$ in the *linear approximation* to the increment in $y = f(x)$. Thus, if Δx is sufficiently small we can write

$$\Delta y = \Delta f = f(x + \Delta x) - f(x) \approx f'(x)\,\Delta x$$

Comparing this with what we would like, $\Delta z \approx f_x\,\Delta x + f_y\,\Delta y$, if $f(x, y)$ is differentiable, it now becomes clear that we can achieve this by generalizing the form of the definition of the differentiability given in equation (5.1).

● *Differentiability: functions of two variables* ─────────

Let f be a real-valued function of two variables that is defined in a neighbourhood of a point (x, y) and such that $f_x(x, y)$ and $f_y(x, y)$ exist. Then f is *differentiable* at (x, y) if there exist functions $\varepsilon_1(\Delta x, \Delta y)$ and $\varepsilon_2(\Delta x, \Delta y)$ such that

$$\Delta f = f(x + \Delta x, y + \Delta y) - f(x, y)$$
$$= f_x(x, y)\Delta x + f_y(x, y)\Delta y + \varepsilon_1(\Delta x, \Delta y)\Delta x + \varepsilon_2(\Delta x, \Delta y)\Delta y \qquad (5.2)$$

where

$$\lim_{(\Delta x, \Delta y) \to (0, 0)} \varepsilon_1(\Delta x, \Delta y) = 0$$

$$\lim_{(\Delta x, \Delta y) \to (0, 0)} \varepsilon_2(\Delta x, \Delta y) = 0$$

Clearly, this is the natural linear generalization of definition (5.1). It is also clear how to generalize it for functions of more variables.

PROBLEM 3

> Generalize the above definition for a function of three variables.

If f is a real-valued function of three variables that is defined in the neighbourhood of a point (x, y, z) and such that $f_x(x, y, z)$, $f_y(x, y, z)$ and $f_z(x, y, z)$ exist, then f is *differentiable* at (x, y, z) if there exist functions $\varepsilon_1(\Delta x, \Delta y, \Delta z)$, $\varepsilon_2(\Delta x, \Delta y, \Delta z)$, $\varepsilon_3(\Delta x, \Delta y, \Delta z)$, such that

$$\Delta f = f(x + \Delta x, y + \Delta y, z + \Delta z) - f(x, y, z)$$

$$= f_x(x, y, z)\Delta x + f_y(x, y, z)\Delta y + f_z(x, y, z)\Delta z$$

$$+ \varepsilon_1(\Delta x, \Delta y, \Delta z)\Delta x + \varepsilon_2(\Delta x, \Delta y, \Delta z)\Delta y + \varepsilon_3(\Delta x, \Delta y, \Delta z)\Delta z$$

where

$$\lim_{(\Delta x, \Delta y, \Delta z) \to (0, 0, 0)} \varepsilon_i(\Delta x, \Delta y, \Delta z) = 0 \qquad i = 1, 2, 3$$

The pattern is obvious, and the extension to $\mathbb{R}^n$ presents no difficulties. You are probably wondering, however, why I have not jumped straight into vector notation at this point. I am leaving this to Chapter 9, since the introduction of vectors brings a whole new perspective to the subject.

The result (4.1) (Section 4.6) for the increment Δz in $z = f(x, y)$ can now be easily derived for differentiable functions, from the above definition of differentiability. All we have to do is utilize the fact that since ε_1, $\varepsilon_2 \to 0$ in (5.2) we can choose them sufficiently small that $\varepsilon_1 \Delta x$ and $\varepsilon_2 \Delta y$ may be neglected, leaving the approximation (4.1) as a result.

PROBLEM 4

> Show that $f(x, y) = x^2 + y^2$ is differentiable in $\mathbb{R}^2$.

From Problem 10 of Chapter 4 we have

$$\Delta f = 2x\Delta x + 2y\Delta y + (\Delta x)^2 + (\Delta y)^2$$

$$\equiv f_x \Delta x + f_y \Delta y + \Delta x . \Delta x + \Delta y . \Delta y$$

Now put $\varepsilon_1(\Delta x, \Delta y) = \Delta x$ and $\varepsilon_2(\Delta x, \Delta y) = \Delta y$, and we get

$$\Delta f = f_x \Delta x + f_y \Delta y + \varepsilon_1(\Delta x, \Delta y)\Delta x + \varepsilon_2(\Delta x, \Delta y)\Delta y$$

where

$$\lim_{(\Delta x, \Delta y) \to (0, 0)} \varepsilon_1(\Delta x, \Delta y) = \lim_{(\Delta x, \Delta y) \to (0, 0)} \Delta x = 0$$

and similarly

$$\lim_{(\Delta x, \Delta y)\to(0,0)} \varepsilon_2(\Delta x, \Delta y) = \lim_{(\Delta x, \Delta y)\to(0,0)} \Delta y = 0$$

This proves that $f(x, y)$ is differentiable.

Now, remember that we got into all this by noting that the existence alone of partial derivatives did not imply the sort of differentiability that implies continuity. So does the definition of differentiability given above imply continuity? Yes:

● Theorem: Differentiability implies continuity ───────

If f is differentiable at (x_0, y_0) then it is continuous at (x_0, y_0).

PROOF
To prove continuity, we know from Section 3.3 that we have to ensure that f is defined at (x_0, y_0) and that $\lim_{(x, y)\to(x_0, y_0)} f(x, y)$ exists and is equal to $f(x_0, y_0)$.
Certainly f is defined at (x_0, y_0), so we just have to show that

$$\lim_{(x, y)\to(x_0, y_0)} f(x, y) = f(x_0, y_0)$$

Define $\Delta x = x - x_0$, $\Delta y = y - y_0$; then we have to show that

$$\lim_{(\Delta x, \Delta y)\to(0,0)} f(x_0 + \Delta x, y_0 + \Delta y) = f(x_0, y_0)$$

Now since f is differentiable at (x_0, y_0) we have

$$f(x_0 + \Delta x, y_0 + \Delta y) - f(x_0, y_0) = f_x(x_0, y_0)\Delta x + f_y(x_0, y_0)\Delta y$$
$$+ \varepsilon_1(\Delta x, \Delta y)\Delta x + \varepsilon_2(\Delta x, \Delta y)\Delta y$$

where

$$\lim_{(\Delta x, \Delta y)\to(0,0)} \varepsilon_i(\Delta x, \Delta y) = 0 \qquad i = 1, 2$$

So clearly

$$\lim_{(\Delta x, \Delta y)\to(0,0)} [f(x_0 + \Delta x, y_0 + \Delta y) - f(x_0, y_0)] = 0$$

and therefore

$$\lim_{(\Delta x, \Delta y)\to(0,0)} f(x_0 + \Delta x, y_0 + \Delta y) = f(x_0, y_0)$$

as required.

Note that the converse of the above result is not true in general – for example, $f(x, y) = \sqrt{(x^2 + y^2)}$ is continuous everywhere, but is not differentiable at the origin, since, in particular, $f_x(0, 0)$ does not exist (*see* Exercises on 5.3).

EXERCISES ON 5.2

1. Generalize the definition of differentiability to functions defined on $\mathbb{R}^n$.

2. Show from first principles that the functions (i) $x + y$, (ii) xy, (iii) $x \cos y$ and (iv) $e^{xy}\cos(x + y)$ are each differentiable in $\mathbb{R}^2$.

3. Discuss the differentiability of the function

$$f(x, y) = \begin{cases} \dfrac{3xy + y^3}{x^2 + y^2} & (x, y) \neq (0, 0) \\ 0 & (x, y) = (0, 0) \end{cases}$$

(*see* Exercises on 3.3).

5.3 Conditions for a function to be differentiable

So (5.2) does indeed provide us with a reasonable definition of differentiability. But we have already seen that the existence of f_x, f_y does not fulfil this definition – so what *is* needed to do so, i.e. what sorts of functions *are* differentiable? It turns out that we have to demand not only the *existence* of f, f_x, f_y but also their *continuity*.

● *Theorem: Conditions for f to be differentiable* ────────────

Let f, f_x, f_y be defined and continuous in a neighbourhood of (x, y). Then f is differentiable at (x, y). If these conditions are met, we say f is *continuously differentiable* or *smooth* at (x, y).

PROOF
This is quite a trek, but it is important. It relies on the mean value theorem for a function of a single variable:

Mean value theorem. Let f be continuous on $[a, b]$ and differentiable on (a, b). Then there is a number c in (a, b) such that

$$f(b) - f(a) = f'(c)(b - a)$$

Kopp (1996, Section 9.1) gives a discussion and proof.

The core of our proof is to apply the mean value theorem to the x- and y-variables separately.

First, since f, f_x, f_y are all continuous in a neighbourhood N of (x, y) we can choose $\Delta x, \Delta y$ sufficiently small that $(x + \Delta x, y + \Delta y)$ lies in N. Then

$$\Delta f(x, y) = f(x + \Delta x, y + \Delta y) - f(x, y)$$

$$= [f(x + \Delta x, y + \Delta y) - f(x + \Delta x, y)] + [f(x + \Delta x, y) - f(x, y)] \qquad \text{(i)}$$

For fixed $x + \Delta x$, $f(x + \Delta x, y)$ is a function of y that is continuous and differentiable in the interval $[y, y + \Delta y]$. So, by the mean value theorem, there is a number c_2 between y and $y + \Delta y$ such that

$$f(x + \Delta x, y + \Delta y) - f(x + \Delta x, y) = f_y(x + \Delta x, c_2)[(y + \Delta y) - y]$$

$$= f_y(x + \Delta x, c_2)\Delta y$$

Note carefully that this is the *single*-variable mean value theorem in action.

Similarly, for fixed y, $f(x, y)$ is a function of x only, and we get

$$f(x + \Delta x, y) - f(x, y) = f_x(c_1, y)\Delta x$$

where c_1 is between x and $x + \Delta x$. So, (i) may be written

$$\Delta f(x, y) = f_x(c_1, y)\Delta x + f_y(x + \Delta x, c_2)\Delta y \qquad (ii)$$

but of course we do not know c_1 and c_2. However, since f_x and f_y are both continuous at (x, y), then since c_1 is between x and $x + \Delta x$ and c_2 is between y and $y + \Delta y$, we have

$$\lim_{(\Delta x, \Delta y) \to (0, 0)} f_x(c_1, y) = f_x(x, y)$$

and

$$\lim_{(\Delta x, \Delta y) \to (0, 0)} f_y(x + \Delta x, c_2) = f_y(x, y)$$

Now put

$$\varepsilon_1(\Delta x, \Delta y) = f_x(c_1, y) - f_x(x, y)$$

It follows that

$$\lim_{(\Delta x, \Delta y) \to (0, 0)} \varepsilon_1(\Delta x, \Delta y) = 0$$

Similarly, if

$$\varepsilon_2(\Delta x, \Delta y) = f_y(x + \Delta x, c_2) - f_y(x, y)$$

then

$$\lim_{(\Delta x, \Delta y) \to (0, 0)} \varepsilon_2(\Delta x, \Delta y) = 0$$

We can now write (ii) as

$$\Delta f(x, y) = f(x + \Delta x, y + \Delta y) - f(x, y)$$
$$= (f_x(x, y) + \varepsilon_1(\Delta x, \Delta y))\Delta x + (f_y(x, y) + \varepsilon_2(\Delta x, \Delta y))\Delta y$$
$$= f_x(x, y)\Delta x + f_y(x, y)\Delta y + \varepsilon_1(\Delta x, \Delta y)\Delta x + \varepsilon_2(\Delta x, \Delta y)\Delta y$$

which finally proves differentiability.

Note that the converse of this theorem is not true – if f is differentiable, it does not follow that f_x, f_y are continuous – *see*, for example, Further Exercises, Question 2.

EXERCISES ON 5.3

1. Show that the following functions are differentiable:
   (i) $f(x, y) = x \cos(x + y)$ (ii) $f(x, y) = xe^{x+y^2}$

   (iii) $\left(\sum_{i=1}^{5} x_i^2 \right) \exp\left(\sum_{j=1}^{5} x_j \right)$

2. (i) If $f(x, y) = \sqrt{x^2 + y^2}$, express $f_x(0, 0)$ as a limit and hence deduce that it does not exist.

   (ii) Show that $f(x, y)$ is continuous at $(0, 0)$ and comment on your results.

3. Discuss the differentiability of the function

$$f(x, y, z) = \frac{e^{x+y+z}}{\sqrt{x^2 + y^2 + z^2}}$$

at the origin.

5.4 Proof of the chain rules

Here we prove the chain rules given in Section 4.5, using the notion of differentiability given in this chapter. As noted in Section 4.5, we only have to prove chain rule 2, since rule 1 follows from this.

Since x and y are both functions of r and s, a change Δr in r will produce a change Δx in x and Δy in y. So we can write:

$$\Delta x = x(r + \Delta r, s) - x(r, s)$$

$$\Delta y = y(r + \Delta r, s) - y(r, s)$$

As z is differentiable, we can write

$$\Delta z = \frac{\partial z}{\partial x} \Delta x + \frac{\partial z}{\partial y} \Delta y + \varepsilon_1(\Delta x, \Delta y)\Delta x + \varepsilon_2(\Delta x, \Delta y)\Delta y$$

where

$$\lim_{(\Delta x, \Delta y) \to (0, 0)} \varepsilon_1(\Delta x, \Delta y) = \lim_{(\Delta x, \Delta y) \to (0, 0)} \varepsilon_2(\Delta x, \Delta y) = 0$$

Dividing both sides of the expression for Δz by Δr and taking limits gives

$$\lim_{\Delta r \to 0} \frac{\Delta z}{\Delta r} = \lim_{\Delta r \to 0} \left[\frac{\partial z}{\partial x}\frac{\Delta x}{\Delta r} + \frac{\partial z}{\partial y}\frac{\Delta y}{\Delta r} + \varepsilon_1(\Delta x, \Delta y)\frac{\Delta x}{\Delta r} + \varepsilon_2(\Delta x, \Delta y)\frac{\Delta y}{\Delta r} \right]$$

Now since x and y are continuous functions of r, we have

$$\lim_{\Delta r \to 0} \Delta x = \lim_{\Delta r \to 0} \Delta y = 0$$

and hence

$$\lim_{\Delta r \to 0} \varepsilon_1(\Delta x, \Delta y) = \lim_{(\Delta x, \Delta y) \to (0, 0)} \varepsilon_1(\Delta x, \Delta y) = 0$$

Similarly

$$\lim_{\Delta r \to 0} \varepsilon_2(\Delta x, \Delta y) = \lim_{(\Delta x, \Delta y)(0, 0)} \varepsilon_2(\Delta x, \Delta y) = 0$$

So

$$\lim_{\Delta r \to 0} \frac{\Delta z}{\Delta r} = \frac{\partial z}{\partial r} = \frac{\partial z}{\partial x}\frac{\partial x}{\partial r} + \frac{\partial z}{\partial y}\frac{\partial y}{\partial r} + 0 \cdot \frac{\partial x}{\partial r} + 0 \cdot \frac{\partial y}{\partial r}$$

and the chain rule follows.

Extend this proof to the case of functions of three variables.

5.5 The tangent plane as a linear approximation to a surface

We now look at how the condition of differentiability leads naturally to the tangent plane view of partial differentiation.

Suppose $z = f(x, y)$ is differentiable at a point $(x_0, y_0, z_0 = f(x_0, y_0))$. Now if (x, y) is a point sufficiently close to (x_0, y_0) we can write, taking $\Delta x = x - x_0$, $\Delta y = y - y_0$, $\Delta z = z - z_0$, and using the definition of differentiability (5.2),

$$\Delta z = z - z_0 = f_x(x_0, y_0)(x - x_0) + f_y(x_0, y_0)(y - y_0)$$

This is the equation of a plane through the point (x_0, y_0, z_0):

$$f_x(x_0, y_0)(x - x_0) + f_y(x_0, y_0)(y - y_0) - (z - z_0) = 0$$

This plane is perpendicular to the vector

$$f_x(x_0, y_0)\mathbf{i} + f_y(x_0, y_0)\mathbf{j} - \mathbf{k}$$

through (x_0, y_0, z_0). It is called the *tangent plane to the surface* $z = f(x, y)$ *at* (x_0, y_0, z_0). We have obtained it by using differentiability of $f(x, y)$ to give us a linear approximation to the surface at (x_0, y_0, z_0). We will say more about tangent planes in Chapter 9. Here we are simply treating it as the linear approximation which is permissible when a function is differentiable.

PROBLEM 5

Find the equation of the tangent plane to the surface $z = x^2y + y^2$ at the point $(1, 1, 2)$. Find the normal to the surface at this point.

For $z = f(x, y) = x^2y + y^2$ we have

$$f_x = 2xy \qquad\qquad f_y = x^2 + 2y$$

So

$$f_x(1, 1) = 2 \qquad\qquad f_y(1, 1) = 3$$

and the equation of the tangent plane is thus

$$z - 2 = f_x(1, 1)(x - 1) + f_y(1, 1)(y - 1)$$

or

$$z - 2 = 2(x - 1) + 3(y - 1)$$

i.e.

$$2x + 3y - z = 3$$

The normal to this plane (and therefore to the surface) at $(1, 1, 2)$ is $2\mathbf{i} + 3\mathbf{j} - \mathbf{k}$.

We referred above to the 'linear approximation' to the surface – let us see what this actually means by considering the case of a function of a single variable in detail.

Consider a function $y = f(x)$, and suppose we increase x to $x + \Delta x$, producing an *increment* in y, $\Delta y = f(x + \Delta x) - f(x)$. This is illustrated in Fig. 5.1. Now, in general, $f(x + \Delta x)$ may be difficult to calculate and so we may not necessarily know Δy, even though the above result gives it exactly. Can we find an approximation to Δy, which will suffice for purposes when Δx is 'small'? The mean value theorem (Section 5.3) provides a hint. Putting $b = x + \Delta x$, $a = x$, $c = \zeta$ tells us that, provided $f(x)$ is differentiable over the interval under consideration, then there is a point ζ between x and $x + \Delta x$ such that

$$\Delta y = f(x + \Delta x) - f(x) = f'(\zeta)\Delta x$$

The trouble is, we do not know ζ. It is in fact the value of x at the point on the curve for which the tangent is parallel to the chord MN in Fig. 5.1. We have no easy way of finding such a ζ. However, we can get a good approximation to the chord MN by using the tangent, MP, to the curve at x. Its slope is $f'(x)$ and using this we can calculate PL, which, for small Δx, is approximately equal to NL. We can therefore write

$$\Delta y \approx f'(x)\Delta x$$

This is a *linear approximation* in Δx. It is equivalent graphically to replacing the curve between M and N by the tangent line segment to the curve at M. It is called the *linearization of $f(x)$*. In the mean value theorem it essentially replaces $f'(\zeta)$, which we do not know, by $f'(x)$, which we do. This linearization depends on $f(x)$ being differentiable at x.

In general, we say that if $y = f(x)$ is differentiable at $x = a$, then

$$L(x) = f(a) + f'(a)(x - a)$$

is the *linearization of f at a*. The approximation

$$f(x) \approx L(x)$$

is the *standard linear approximation* of f at a.

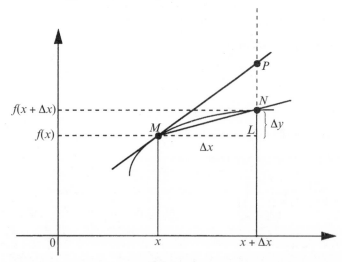

Fig. 5.1

Although we have approached this concept by looking at the graph of $f(x)$, it is not dependent on geometry and can be made analytically rigorous, by using the mean value theorem.

So much for functions of a single variable. We now look at how all this can be generalized to the case of functions of two variables. The crucial question comes down to what happens if we increase x and y by 'infinitesimal' *increments* Δx and Δy – by how much does $z = f(x, y)$ increase? Now, of course, the exact answer is

$$\Delta z = f(x + \Delta x, y + \Delta y) - f(x, y)$$

This is called the *increment* in z. In principle it can be calculated exactly.

However, we can get a linear approximation to it by using the definition of differentiability. Thus we define the *linearization* of a differentiable function $z = f(x, y)$ at a point (x_0, y_0) as the function

$$L(x, y) = f(x_0, y_0) + f_x(x_0, y_0)(x - x_0) + f_y(x_0, y_0)(y - y_0)$$

The approximation

$$f(x, y) \simeq L(x, y)$$

is called the *standard linear approximation* of f at (x_0, y_0). In view of our earlier discussion, it is also called the *tangent plane approximation*.

PROBLEM 6

Find the linearization of $f(x, y) = x^2 + x \cos(x + y)$ at the point $(0, \pi)$.

$$f_x(x, y) = 2x + \cos(x + y) - x \sin(x + y)$$
$$f_y(x, y) = - \sin(x + y)$$

So

$$f(0, \pi) = 0$$
$$f_x(0, \pi) = -1$$
$$f_y(0, \pi) = 0$$

The linearization is thus

$$L(x, y) = f(0, \pi) + f_x(0, \pi)(x - 0) + f_y(0, \pi)(y - 0)$$
$$= 0 + (-1)(x) + (0)(y)$$
$$= -x$$

The point about the linearization is that it gives a 'first', and hopefully good, approximation to the function near to the point (x_0, y_0). Thus, in Problem 6, provided (x, y) is 'very close' to $(0, \pi)$ we can replace the complicated $x^2 + x \cos(x + y)$ by the much simpler $-x$.

The linear approximation is actually the linear order term in the Taylor series for the function about (x_0, y_0), which we discuss in the next chapter. This is exactly analogous to how $f(a) + f'(a)(x - a)$ is the linear part of the Taylor series for $f(x)$ about $x = a$.

EXERCISES ON 5.5

1. Find the equation of the tangent plane to the following surfaces at the points given:
   (i) $f(x, y) = 9x^2 + 4y^2$ at $(1, 2, 25)$
   (ii) $f(x, y) = xy \cos(x^2 y^2)$ at $(0, 1, 0)$
2. Find the linearization of each of the functions in Exercise 1.

FURTHER EXERCISES

1. Discuss the differentiability of the following functions:
   (i) $f(x, y) = x^2 y^3$ (ii) $f(x, y) = \dfrac{x + y}{x - y}$

   (iii) $f(x, y, z) = \dfrac{1}{\sqrt{x^2 + y^2 + z^2}}$ (iv) $f(x, y) = \sin(e^{x+y})$

   (v)
   $$f(x, y) = \begin{cases} \dfrac{x + y}{x^2 + y^2} & (x, y) \ne (0, 0) \\ 1 & (x, y) = (0, 0) \end{cases}$$

2. If
   $$f(x, y) = \begin{cases} (x^2 + y^2)\sin\left(\dfrac{1}{\sqrt{x^2 + y^2}}\right) & (x, y) \ne (0, 0) \\ 0 & (x, y) = (0, 0) \end{cases}$$
   (i) show that f_x, f_y are not continuous at $(0, 0)$
   (ii) show that f is differentiable at $(0, 0)$.

3. Show that any polynomial function of x and y is differentiable at any point in $\mathbb{R}^2$.

4. Find the equations for the tangent plane and a vector normal to the surface in the following cases:
   (i) $z = 2x^3 y^3 + 3x^2 y$ at $(-1, 1, 1)$
   (ii) $z = \dfrac{x + y}{x^3}$ at $(1, 0, 1)$
   (iii) $z = (x + y)e^{-xy}$ at $(0, 2, 2)$
   (iv) $z = x^2 + y^2$ at $(0, 0, 0)$
   (v) $z = e^{x+y} \cos(x^2 + y^2)$ at $(0, 0, 1)$
   (vi) $z = x^{-1/2} + y^{-1/3}$ at $(1, 1, 2)$

5. Find all points on the following surfaces at which the tangent plane is horizontal:
   (i) $z = 4x^2 y^3$ (ii) $z = x^2 - 3xy + y^2 - 2x + 3y$

6. Find a point on the surface $z = x^2 - 3y^2$ at which the tangent plane is parallel to the plane $4x + 3y - z = 6$.

7. Show that the surfaces $z = \sqrt{16 - x^2 - y^2}$ and $z = \sqrt{x^2 + y^2}$ intersect at points (a, b) such that $a^2 + b^2 = 8$. By finding the tangent planes at such points show that the surfaces are always perpendicular to each other at the point of intersection.

8. Find the linearization of the following functions at the given points:
   (i) $f(x, y) = x^2 + y^2 + 1$ at (a) $(0, 0)$ (b) $(1, -1)$
   (ii) $f(x, y) = 4x - 3y + 2$ at (a) $(0, 0)$ (b) $(1, 1)$
   (iii) $f(x, y) = x^3 y^5$ at (a) $(0, 0)$ (b) $(2, 1)$
   (iv) $f(x, y) = e^{2x} \sin y$ (a) $(0, 0)$ (b) $(0, \pi/2)$

9. Find the linearization of the following functions at the given points:
   (i) $f(x, y, z) = xy + yz + xz$ at (a) $(0, 0, 0)$ (b) $(1, 0, 0)$ (c) $(-1, 1, 0)$
   (ii) $f(x, y, z) = x^2 + y^2 + z^2$ at (a) $(0, 0, 0)$ (b) $(0, 1, 0)$ (c) $(1, 1, 0)$

10. Essay or discussion topics:
    (i) Interpretations of f_x, f_y.
    (ii) The connection between differentiability and continuity for functions of two variables.
    (iii) The role of the mean value theorem in differentiability of functions of two variables.
    (iv) Differentiability and the chain rules.
    (v) Linearization and differentiability.

6 • Taylor Series for Functions of Several Variables

6.1 Introduction

PROBLEM 1

Give two definitions each of (i) $\sin x$, (ii) e^x

(i) You probably gave the trigonometric definition of $\sin x$

$$\sin x = \frac{\text{Opposite}}{\text{Hypotenuse}}$$

and the series definition

$$\sin x = x - \frac{x^3}{3!} + \frac{x^5}{5!} - \ldots$$

(ii) For e^x you might have given the limit definition

$$e^x = \lim_{n \to \infty} \left(1 + \frac{x}{n}\right)^n$$

and again you probably gave the series definition

$$e^x = 1 + x + \frac{x^2}{2!} + \frac{x^3}{3!} + \ldots$$

These examples are enough to show the utility and power of the series definition for functions of a single variable. Such definitions are useful from the theoretical point of view because they provide a uniform means of expressing *any* function in terms of sums (albeit infinite sums) of powers. They are important from a practical point of view because they provide a means of evaluating a function to any required accuracy. The key point about such a series (technically, a Maclaurin series – or Taylor series about the origin $x = 0$) is that the only terms appearing are powers of x. In general,

$$f(x) = a_0 + a_1 x + a_2 x^2 + \ldots + a_r x^r + \ldots$$

PROBLEM 2

Write down the typical term you might expect in a series expansion of a function of two variables $f(x, y)$.

It does not take much imagination to guess that you will get terms like $x^r y^s$ for all integer values of r and s. This would be for 'expansion about the origin'. A general series might start something like

$$f(x, y) = a_{00} + a_{10}x + a_{01}y + a_{20}x^2 + a_{11}xy + a_{02}y^2 + \dots$$

The pattern is fairly obvious – but how would we find the coefficients a_{ij}? Remember that in the case of a function of a single variable the coefficients are given by differentiation – for example (*see* Pearson, 1996),

$$a_n = \frac{f^{(n)}(0)}{n!}$$

So you can bet that the a_{ij} are likewise going to involve derivatives, but partial derivatives.

You may also remember that in the rigorous formal study of Taylor and Maclaurin series of a single variable we avoided the sloppiness of the '+ …' at the end of the series by using a *remainder term* which tended to zero as we took more and more terms of the series (*see* Kopp, 1996). We use a similar device in the theory of the Taylor series of functions of more than one variable.

Taylor series are not only useful for evaluating functions – they can tell us a lot about the general behaviour of a function near special points such as *extrema*, i.e. maxima or minima points. Thus, suppose we have a function

$$f(x) = f(0) + f'(0)x + \frac{f''(0)}{2}x^2 + \dots$$

which has a turning point at the origin, $x = 0$. Then we know that $f'(0) = 0$ and so the function has the form

$$f(x) = f(0) + \frac{f''(0)}{2}x^2 + \dots$$

Now very close to the origin we can neglect powers of x greater than the second and write

$$f(x) \approx f(0) + \frac{f''(0)}{2!}x^2$$

So the function behaves like a quadratic near the origin. The nature of the turning point is then determined from the sign of $f''(0)$. If $f''(0) > 0$ then we have a downward-pointing parabola and therefore have a minimum. If $f''(0) < 0$ then we have a maximum. If $f''(0) = 0$ then we have to look at higher-order terms. The point here is that we can deduce information about stationary values *from the form of the Taylor series*. It will come as no surprise to you that we can do exactly the same thing for functions of several variables. Indeed, this is one of the main applications of such Taylor series.

EXERCISE ON 6.1

Obtain power series expansions about the origin for the functions (i) e^{xy}, (ii) $y\cos(xy)$, (iii) $\sin(x + y)$.

6.2 Taylor series for functions of two variables

This is one of those topics where the details and the complications of the results belie their essential simplicity and structure. The results look messy, but provided you are at home with the Taylor series for a single variable and with the binomial theorem, then things are not too bad.

Before we launch into the formal theory of power series for functions of two variables, let us see what we can do using what we know about series for functions of a single variable. The important point to note is that the power series for a sufficiently well-behaved function of x, y is unique. That is, if $f(x, y)$ is continuously differentiable to all orders of derivative and if

$$f(x, y) = \sum_{\substack{r=0 \\ s=0}}^{\infty} a_{rs} x^r y^s = \sum_{\substack{r=0 \\ s=0}}^{\infty} b_{rs} x^r y^s$$

then

$$a_{rs} = b_{rs} \text{ for all } r, s$$

So if we obtain such a series, by whatever means, then that must be the required Taylor series. We can often do this using elementary series for a function of a single variable. For example,

$$x \cos(xy) = x\left(1 - \frac{(xy)^2}{2!} + \frac{(xy)^4}{4!} - \frac{(xy)^6}{6!} + \dots\right)$$

$$= x - \frac{x^3 y^2}{2} + \frac{x^5 y^4}{24} - \frac{x^7 y^6}{720} + \dots$$

We now turn to the formal theory of power series for functions of two variables. As often in the theory of functions of several variables, we start from the analogous case in a single variable and generalize that in some way. So let us start with the following.

● Taylor's theorem for a function of a single variable ─────────

If $f(x)$ is a function of a single variable x which has continuous derivatives up to $(n + 1)$th order in an interval I, then, for any numbers x, $x + t$ on this interval,

$$f(x + t) = f(x) + tf'(x) + \frac{t^2}{2!} f''(x) + \frac{t^3}{3!} f'''(x) + \dots$$

$$+ \frac{t^n}{n!} f^{(n)}(x) + R_n$$

where the remainder term, R_n, may be written in the (*Lagrange*) form

$$R_n = \frac{t^{n+1}}{(n+1)!} f^{(n+1)}(x + \theta t) \qquad 0 < \theta < 1$$

The special case when we take $x = 0$ yields *Maclaurin's theorem*:

$$f(t) = f(0) + tf'(0) + \frac{t^2}{2!} f''(0) + \frac{t^3}{3!} f'''(0) + \dots$$

$$+ \frac{t^n}{n!} f^{(n)}(0) + R_n$$

where

$$R_n = \frac{t^{n+1}}{(n+1)!} f^{(n+1)}(\theta t) \qquad 0 < \theta < 1$$

which is also referred to as *expansion about the origin*.

Now consider a function of two variables $f(x, y)$ whose partial derivatives up to order $n + 1$ are all continuous in some region R of the xy-plane. Here comes the clever bit. Define a function

$$F(t) = f(x + ht, y + kt) \equiv f(\alpha, \beta)$$

treated as a function of t alone, for x, y, h, k all regarded as parameters. Clearly $F(0) = f(x, y)$. Also, by the conditions imposed on $f(x, y)$, $F(t)$ is a continuous function of t whenever $(x + ht, y + kt)$ lies in the region R. Also, $F'(t)$ exists and is continuous on R.

PROBLEM 3

Show that

$$F'(t) = h\frac{\partial f}{\partial x} + k\frac{\partial f}{\partial y}$$

and obtain an expression for $F^{(n)}(t)$.

We only have to apply the expression for the total derivative :

$$F'(t) = \frac{\partial f}{\partial \alpha}\frac{d\alpha}{dt} + \frac{\partial f}{\partial \beta}\frac{d\beta}{dt} = h\frac{\partial f}{\partial \alpha} + k\frac{\partial f}{\partial \beta}$$

$$= h\frac{\partial f}{\partial x} + k\frac{\partial f}{\partial y}$$

We can think of this in the *operator* form

$$F'(t) = \left(h\frac{\partial}{\partial x} + k\frac{\partial}{\partial y}\right) f(x + ht, y + kt)$$

i.e. the operator

$$D = h\frac{\partial}{\partial x} + k\frac{\partial}{\partial y}$$

operating on f:

$$Df = \left(h\frac{\partial}{\partial x} + k\frac{\partial}{\partial y}\right) f(x + ht, y + kt) = h\frac{\partial f}{\partial x} + k\frac{\partial f}{\partial y}$$

This operator notation is very suggestive, and using it we find that we can write, for example,

$$F''(t) = D^2 f = \left(h\frac{\partial}{\partial x} + k\frac{\partial}{\partial y}\right)^2 f(x + ht, y + kt)$$

$$= h^2\frac{\partial^2 f}{\partial x^2} + 2hk\frac{\partial^2 f}{\partial x\partial y} + k^2\frac{\partial^2 f}{\partial y^2}$$

Furthermore, the nth derivative may be written:

$$F^{(n)}(t) = \left(h\frac{\partial}{\partial x} + k\frac{\partial}{\partial y}\right)^n f(x + ht, y + kt)$$

Continuity of $F(t)$, $F'(t)$, ..., $F^{(n+1)}(t)$ follows from the continuity of all partial derivatives of f. We can therefore apply Maclaurin's theorem for a single variable to $F(t)$ and get

$$F(t) = F(0) + tF'(0) + \frac{t^2}{2!}F''(0) + \frac{t^3}{3!}F'''(0) + \dots + \frac{t^n}{n!}F^n(0) + R_n$$

where

$$R_n = \frac{t^{n+1}}{(n+1)!}F^{(n+1)}(\theta t) \qquad 0 < \theta < 1$$

If we now return to the original (x, y) variables and put $t = 1$ then this finally gives us:

● Taylor's theorem for a function of two variables ───────

For simplicity, we state the theorem for a rectangular region. If $f(x, y)$ and its partial derivatives up to order $n + 1$ are continuous throughout an open rectangular region R centred on (x, y), then, throughout R,

$$f(x + h, y + k) = f(x, y) + \left(h\frac{\partial}{\partial x} + k\frac{\partial}{\partial y}\right)f(x, y) + \frac{1}{2!}\left(h\frac{\partial}{\partial x} + k\frac{\partial}{\partial y}\right)^2 f(x, y) + \dots$$

$$+ \frac{1}{n!}\left(h\frac{\partial}{\partial x} + k\frac{\partial}{\partial y}\right)^n f(x, y) + R_n$$

where

$$R_n = \frac{1}{(n+1)!}\left(h\frac{\partial}{\partial x} + k\frac{\partial}{\partial y}\right)^{n+1} f(x + \theta h, y + \theta k) \qquad 0 < \theta < 1$$

Note that this is very similar to Taylor's theorem for a single variable, with $f^{(n)}(x) = D^n f(x)$ in which $D = d/dx$ is replaced by the operator $D \equiv h\partial/\partial x + k\partial/\partial y$.
You may have a bit of trouble with expressions like D^n for $n > 1$, but you only have to think of the binomial theorem, as indicated in the next problem.

PROBLEM 4

Write out

$$D^3 f \equiv \left(h\frac{\partial}{\partial x} + k\frac{\partial}{\partial y}\right)^3 f$$

in full.

Simply imagine $h\partial/\partial x$, $k\partial/\partial y$ as terms in a binomial expression and 'expand' by the binomial theorem to get

$$D^3 f = \left[h^3 \left(\frac{\partial}{\partial x}\right)^3 + 3h^2 \left(\frac{\partial}{\partial x}\right)^2 \left(k\frac{\partial}{\partial y}\right) + 3\left(h\frac{\partial}{\partial x}\right)\left(k\frac{\partial}{\partial y}\right)^2 + \left(k\frac{\partial}{\partial y}\right)^3 \right] f$$

$$\equiv h^3 \frac{\partial^3 f}{\partial x^3} + 3h^2 k \frac{\partial^3 f}{\partial x^2 \partial y} + 3hk^2 \frac{\partial^3 f}{\partial x \partial y^2} + k^3 \frac{\partial^3 f}{\partial y^3}$$

Another possible area of confusion is the remainder term, which is evaluated at some point $(x + \theta h, y + \theta k)$ which is unknown, because θ is unknown. The point about this is that we can be sure such a point *exists*, even though we do not know where it is. In practice, we are not usually concerned with the remainder terms – we know that by taking enough terms in the series they can be ignored. We therefore usually write Taylor's theorem in the form of an infinite expansion:

$$f(x + h, y + k) = f(x, y) + \left(h\frac{\partial}{\partial x} + k\frac{\partial}{\partial y}\right) f(x, y) + \frac{1}{2!}\left(h\frac{\partial}{\partial x} + k\frac{\partial}{\partial y}\right)^2 f(x, y) + \ldots$$

$$+ \frac{1}{n!}\left(h\frac{\partial}{\partial x} + k\frac{\partial}{\partial y}\right)^n f(x, y) + \ldots \tag{6.1}$$

PROBLEM 5

Obtain the Taylor series for $f(x, y) = x^2 + y^2$. What can you say about the Taylor series for the nth-degree polynomial in x and y:

$$p(x, y) = a_{n, 0}\, x^n + a_{n-1, 1}\, x^{n-1} y + a_{n-2, 2}\, x^{n-2} y^2 + \ldots$$

$$+ a_{1, n-1}\, xy^{n-1} + a_{0, n}\, y_n?$$

We have

$$\left(h\frac{\partial}{\partial x} + k\frac{\partial}{\partial y}\right)(x^2 + y^2) = h(2x) + k(2y)$$

$$\left(h\frac{\partial}{\partial x} + k\frac{\partial}{\partial y}\right)^2 (x^2 + y^2) = 2h^2 + 2k^2$$

and

$$\left(h\frac{\partial}{\partial x} + k\frac{\partial}{\partial y}\right)^n (x^2 + y^2) = 0 \qquad \text{for } n > 2$$

So the Taylor series terminates at the second-order term and we get

$$f(x + h, y + k) = x^2 + y^2 + 2xh + 2ky + \frac{1}{2!}(2h^2 + 2k^2)$$

$$= x^2 + 2xh + h^2 + y^2 + 2ky + k^2$$

$$\equiv (x + h)^2 + (y + k)^2$$

In other words, the Taylor series is simply the expansion of the polynomial $f(x + h, y + k)$ into terms of the form of products of powers of x, y, h, k. This is always true when $f(x, y)$ is a polynomial in (x, y) – the Taylor series of a polynomial *is* the polynomial. So the Taylor series for $p(x, y)$ is simply the expansion of the terms in $p(x + h, y + k)$.

As you will probably know from your work on functions of a single variable, the Taylor series itself is rarely used. We can almost always manage by using the Taylor series about the origin, i.e. the *Maclaurin series*. In the case of functions of two variables the Maclaurin series may be written in the form

$$f(x, y) = f(0, 0) + \left(x\frac{\partial}{\partial x} + y\frac{\partial}{\partial y}\right)f(0, 0) + \frac{1}{2!}\left(x\frac{\partial}{\partial x} + y\frac{\partial}{\partial y}\right)^2 f(0, 0) + \dots$$

$$+ \frac{1}{n!}\left(x\frac{\partial}{\partial x} + y\frac{\partial}{\partial y}\right)^n f(0, 0) + \dots$$

where, for example,

$$\left(x\frac{\partial}{\partial x} + y\frac{\partial}{\partial y}\right)^2 f(0, 0) \equiv x^2 f_{xx}\Big|_{(0, 0)} + 2xy f_{xy}\Big|_{(0, 0)} + 2y^2 f_{yy}\Big|_{(0, 0)}$$

means that the derivatives of $f(x, y)$ are all evaluated at the origin $(0, 0)$.

Note that it is no real restriction taking the Taylor series about the origin. To obtain the series about any other point we have only to shift the origin by a translation of the x- and y-axes.

EXERCISES ON 6.2

1. Obtain the Maclaurin series up to terms of order 3 for the functions
   (i) e^{x+y} (ii) $x \sin y$ (iii) $\cos(x^2 + y)$
   Compare with other approaches to finding the power series.
2. Obtain the Maclaurin series up to terms of order 4 for the functions
   (i) $x^5 y$ (ii) e^{xy} (iii) $\sin(x^2 + y^2)$

6.3 Taylor's theorem for functions of more than two variables: project

This section is really an exercise in coping with intricate notation, a consolidation of what has gone before, and a brief overview of extensions of Taylor's theorem for functions of any number of variables.

Following the same procedure as in Section 6.2, we can extend Taylor's theorem to three variables as follows:

● Taylor's theorem for three variables ——————————

If $f(x, y, z)$ and its partial derivatives up to order $n + 1$ are continuous throughout an open region, R, centred on (x, y, z), then throughout R

$$f(x + h, y + j, z + k) = f(x, y, z) + \left(h\frac{\partial}{\partial x} + j\frac{\partial}{\partial y} + k\frac{\partial}{\partial z}\right)f(x, y, z)$$

$$+ \frac{1}{2!}\left(h\frac{\partial}{\partial x} + j\frac{\partial}{\partial y} + k\frac{\partial}{\partial z}\right)^2 f(x, y, z)$$

$$+ \dots +$$

$$+ \frac{1}{n!} \left(h\frac{\partial}{\partial x} + j\frac{\partial}{\partial y} + k\frac{\partial}{\partial z} \right)^n f(x, y, z)$$

$$+ R_n$$

where

$$R_n = \frac{1}{(n+1)!} \left(h\frac{\partial}{\partial x} + j\frac{\partial}{\partial y} + k\frac{\partial}{\partial z} \right)^{n+1} f(x + \theta h, y + \theta j, z + \theta k) \quad 0 < \theta < 1$$

Again, notice the structure – it is not as bad as it might first appear. Evaluating the derivative terms is a little more problematical. For example,

$$\left(h\frac{\partial}{\partial x} + j\frac{\partial}{\partial y} + k\frac{\partial}{\partial z} \right)^2 f = h^2\frac{\partial^2 f}{\partial x^2} + j^2\frac{\partial^2 f}{\partial y^2} + k^2\frac{\partial^2 f}{\partial z^2} + 2jk\frac{\partial^2 f}{\partial y \partial z} + 2kh\frac{\partial^2 f}{\partial z \partial x} + 2hj\frac{\partial^2 f}{\partial x \partial y}$$

While we cannot appeal to the binomial theorem in this case, we can still rely on analogy with ordinary algebra – for example, the parallel with the expansion of $(a + b + c)^2$ is obvious. The key thing here is that the use of the *operator notation* here converts *differentiation* into *algebra*.

Just for the record, I will leave you with a statement of Taylor's theorem for a function of m variables, $x_1, x_2, \ldots, x_m$, which should now be little more than an exercise in coping with notation:

$$f(x_1 + h_1, \ldots, x_m + h_m) =$$

$$f(x_1, \ldots, x_m) + \sum_{k=1}^{n} \frac{1}{k!} \left(\sum_{i=1}^{m} h_i \frac{\partial}{\partial x_i} \right)^k f(x_1, \ldots, x_m) + R_n$$

where

$$R_n = \frac{1}{(n+1)!} \left(\sum_{i=1}^{m} h_i \frac{\partial}{\partial x_i} \right)^{n+1} f(x_1 + \theta h_1, \ldots, x_m + \theta h_m) \quad 0 < \theta < 1$$

6.4 Extreme values for functions of two variables

In the theory of functions of a single variable there is a simple geometric way to look at the study of *extreme values* – or *turning points*, as they are usually called in elementary calculus. Firstly, a simple picture helps – *see* Fig. 6.1.

Extreme values are really of two types – maxima and minima. In each case the tangent at the extreme value is horizontal and so the first derivative $f'(x)$ is zero. The solution of the equation

$$f'(x) = 0$$

gives the *stationary values* or *extrema*.

There are then a number of ways of deciding what sort of stationary value we have – a maximum or minimum, or something else. The most elementary observation about a local maximum is that for every point 'sufficiently close' to it, on either

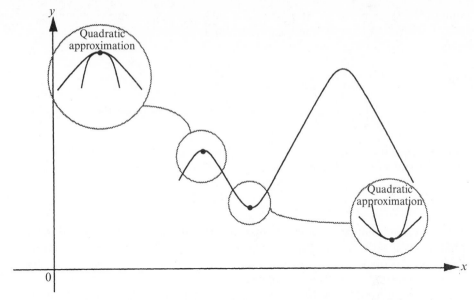

Fig. 6.1

side, the value of $f(x)$ is less than at the maximum. Similarly at every point sufficiently close to a local minimum the value of $f(x)$ is greater than at the minimum.

Another approach, using calculus, is to note that as we pass through a maximum
the gradient $f'(x)$ goes from positive, through zero, to negative. It is therefore
decreasing – its rate of change is therefore negative and so $(f'(x))' = f''(x) < 0$ for a
maximum. Similar arguments tell us that $f''(x) > 0$ at a minimum.

Neither of these approaches extends very well to functions of two (or more) variables. There is, however, the third approach mentioned in Section 6.1, using the
Taylor series. From Fig. 6.1 we see that sufficiently close to a maximum or
minimum the exact shape of the curve is not really that important – the crucial
thing is that at a maximum we have an upwards-pointing hump and at a minimum
we have a downwards-pointing hump. So, at the actual extreme value and close to
it, it would not matter if we replaced the curve itself by a simple quadratic or parabolic approximation to the curve. The nature of the extremum of this parabolic
approximation would be identical to that of the curve itself. We can get such a
quadratic approximation from the Taylor series about the extremum point. Let us
assume that the extremum in question is at the origin (we can always move it there)
and then, as explained in Section 6.1, we can write

$$f(x) - f(0) \simeq \frac{f''(0)}{2!} x^2$$

for x sufficiently close to the origin.

It is now clear that the nature of the extremum depends on the sign of $f''(0)$, and
the results obtained above regarding $f''(0)$ and the nature of the turning point
follow easily. We can take this further. Suppose $f''(0) = 0$, so that it is not clear
what we have at the stationary value. We might have a point of inflexion. At such a

point the curve simply has a 'kink' in it – not a hump. The simplest polynomial with such a kink is a cubic, and so, close to the stationary value we replace the curve by a cubic approximation obtained from the Taylor series and obtain (provided $f'''(0) \neq 0$ of course)

$$f(x) - f(0) \approx \frac{f'''(0)}{3!} x^3$$

Again, the shape of the curve and the nature of the stationary point are identical to that of the approximation – and that depends on the sign of $f'''(0)$. The point of all this, as noted earlier, is that we can get at information about the nature of the stationary point by looking at the truncated Taylor series.

This is something which could be extended to functions of more than one variable, and so it is the approach we adopt. But first let us generalize the actual definitions of stationary values and turning points to such functions. Extrema have a fairly obvious geometric generalization to functions of two variables, but geometric intuition fails us beyond that. Points of inflexion do not generalize easily, even to functions of two variables – they lead us to *saddle points*, which are not easy to appreciate in mathematical terms. So, do not get too attached to geometric views of what we are doing – use them to make the precise definitions sensible, but focus mainly on the mathematical definitions. Also, remember that, just as in functions of a single variable, we are always referring here to *local* extrema.

Let $z = f(x, y)$ be defined on a region R containing a point (x_0, y_0). Then $f(x_0, y_0)$ is a *local maximum* value of f if

$$f(x_0, y_0) \geq f(x, y)$$

for all values of (x, y) in a neighbourhood centred on (x_0, y_0).

PROBLEM 6

Frame a definition for a local minimum.

$f(x_0, y_0)$ is a *local minimum* value of f if

$$f(x_0, y_0) \leq f(x, y)$$

for all values of (x, y) in a neighbourhood centred on (x_0, y_0).

Geometrically, these definitions boil down to generalizations of the simple 'hump' or 'dip' interpretation in the case of functions of a single variable. At *local extrema* the *tangent plane* is horizontal, and for a maximum all local values lie below the plane, whereas for a minimum they all lie above. Such local extrema are also called *relative extrema*.

Just as in the case of a single variable, the above definitions are not of much use in actually *locating* local extrema. In the case of a single variable the first step is to find the *stationary values* by equating the derivative to zero – this is effectively looking for points at which the tangent to the curve is horizontal. In the case of functions of two variables we are essentially looking for points at which the *tangent plane* is horizontal. This suggests the following first step in looking for local extrema:

● *First derivative test* ————————————————————————

If $f(x, y)$ has a local maximum or minimum at a point (x_0, y_0) inside its domain, and if the partial derivatives f_x, f_y exist at that point, then

$$f_x(x_0, y_0) = f_y(x_0, y_0) = 0$$

Note two crucial points:

- The zero derivatives condition is a *necessary condition* only. If local extrema exist, then the derivatives must be zero. It does not follow that if the derivatives are zero then local extrema exist. That is, the condition is not *sufficient*.
- We have been careful to include the qualification 'if the partial derivatives f_x, f_y exist'. We will come back to this later.

The proof of the first derivative test is essentially a formalization of the fact that at an extremum the tangent to any curve on the surface, through the extremum, will be horizontal. In particular, the curves obtained on the surface by taking planes through the extremum parallel to the *xz*- and *yz*-planes will have zero gradient at the extremum. These gradients are given by f_x and f_y respectively.

So, suppose f has a local maximum at an interior point (x_0, y_0) of its domain. Then $x = x_0$ is an interior point of the domain of the curve $z = f(x, y_0)$ in which the plane $y = y_0$ intersects the surface $z = f(x, y)$. Furthermore, $z = f(x, y_0)$ is a differentiable function of x at $x = x_0$, the derivative being $f_x(x_0, y_0)$. Clearly, the function $f(x, y_0)$ has a local maximum at $x = x_0$ and so therefore the value of the derivative of $z = f(x_0, y_0)$ at $x = x_0$ is zero – i.e. $f_x(x_0, y_0) = 0$. An identical argument using $z = f(x_0, y)$ shows also that $f_y(x_0, y_0) = 0$ and the necessary condition for a local maxima follows.

The proof for a local minimum is left for you to do (Exercise 6.4.1).

The first derivative test assures us that a surface $z = f(x, y)$ has a horizontal tangent plane at a local extremum – provided a tangent plane exists there. It also tells us that $f(x, y)$ can only have an extremum value at (i) interior points where $f_x = f_y = 0$, or (ii) interior points where one or both of f_x, f_y do not exist, or (iii) boundary points of the function's domain. A special name is given to the first two cases. An interior point of the domain of a function $f(x, y)$ where both f_x, f_y are zero or where one of them does not exist is called a *critical point* of f. So, $f(x, y)$ can only take extreme values at critical points or boundary points.

Not all critical points correspond to extrema. In the case of functions of a single variable we can have points of inflexion. These are characterized by the fact that the sign of the rate of change of the function does not change as we go through the critical points. Correspondingly, a differentiable function $f(x, y)$ has a saddle point at a critical point (x_0, y_0) if in every neighbourhood of (x_0, y_0) there are points (x, y) where $f(x, y) > f(x_0, y_0)$ and points where $f(x, y) < f(x_0, y_0)$. The corresponding point, $(x_0, y_0, f(x_0, y_0))$, on the surface $z = f(x, y)$ is called a *saddle point of the surface* (*see* Fig. 6.2).

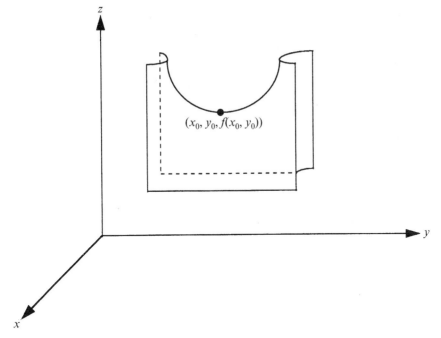

Fig. 6.2

PROBLEM 7

Determine the critical points of the following functions:
(i) $6x - 3x^2 - y^2$
(ii) $x^2 + xy + 3x + 2y + 5$

(i) $f(x, y) = 6x - 3x^2 - y^2$

$$f_x = 6 - 6x \qquad\qquad f_y = -2y$$

The critical points occur at $f_x = f_y = 0$, i.e.

$$6 - 6x = 0$$
$$-2y = 0$$

i.e. at the point $(1, 0)$.

(ii) $f(x, y) = x^2 + xy + 3x + 2y + 5$

$$f_x = 2x + y + 3 = 0 \qquad\qquad f_y = x + 2 = 0$$

The only critical point is thus $(-2, 1)$.

Now all the above test gives us is a necessary condition for a local extremum. It does not tell us for certain that there is an extremum at a critical point, nor what its nature is. We have seen that in the case of functions of a single variable we distinguish maxima from minima by looking at the quadratic term in the Taylor series. We do exactly the same in the case of functions of two variables.

PROBLEM 8

> Suppose $f(x, y)$ and its first and second derivatives are continuous in a neighbourhood of $(x_0 \, y_0)$ and that $f(x, y)$ has a critical point at (x_0, y_0), i.e. $f_x(x_0, y_0) = f_y(x_0, y_0) = 0$. Obtain a quadratic approximation for $f(x, y)$ close to (x_0, y_0).

Using Taylor's theorem (6.1) with $(x, y) = (x_0, y_0)$ we have

$$f(x_0 + h, y_0 + k) = f(x_0, y_0) + hf_x(x_0, y_0) + kf_y(x_0, y_0)$$

$$+ \frac{1}{2}(h^2 f_{xx}(x_0, y_0) + 2hkf_{xy}(x_0, y_0) + k^2 f_{yy}(x_0, y_0)) + \dots$$

So, noting that $f_x(x_0, y_0) = f_y(x_0, y_0) = 0$, and neglecting cubic terms and higher-order terms, we have

$$f(x_0 + h, y_0 + k) \simeq f(x_0, y_0) + \frac{1}{2}(h^2 f_{xx}(x_0, y_0) + 2hkf_{xy}(x_0, y_0) + k^2 f_{yy}(x_0, y_0))$$

So the sign of

$$\Delta f = f(x_0 + h, y + k) - f(x_0, y_0)$$

is determined by the sign of the quadratic expression

$$Q = h^2 f_{xx} + 2hkf_{xy} + k^2 f_{yy}$$

This sort of argument is exactly the one we used in the case of functions of a single variable. Now, however, the second-order derivative term is more complicated. You may remember the fuss made about quadratic functions in your elementary mathematics – here you are going to see one of the reasons why they are so important.

The key difficulty is that of determining the sign of Q for arbitrary values of h and k. This may appear to be a hopeless task – it seems to depend not only on h and k, but also on the second derivatives f_{xx}, f_{xy}, f_{yy}. However, a neat trick simplifies things considerably. Multiply through by f_{xx} and complete the square to get

$$f_{xx}Q = h^2 f_{xx}^2 + 2hkf_{xy}f_{xx} + f_{xx}f_{yy}k^2$$

$$\equiv h^2 f_{xx}^2 + 2hkf_{xy}f_{xx} + k^2 f_{xy}^2 + (f_{xx}f_{yy} - f_{xy}^2)k^2$$

$$\equiv (hf_{xx} + kf_{xy})^2 + (f_{xx}f_{yy} - f_{xy}^2)k^2 \qquad (6.2)$$

Consider the expression

$$D = f_{xx}f_{yy} - f_{xy}^2$$

which (and this is no coincidence) is the *discriminant* of the original quadratic function Q. We can consider three cases, where D is less than, equal to or greater than zero.

First consider the case of $D = f_{xx}f_{yy} - f_{xy}^2 > 0$. The right-hand side of (6.2) is clearly always positive. So, if $f_{xx} > 0$ then, for sufficiently small values of h, k, $Q > 0$ and so we have a local *minimum*. But if $f_{xx} < 0$, then for sufficiently small values of h and k, $Q < 0$ and we have a local *maximum*.

Now consider the case $D = f_{xx}f_{yy} - f_{xy}^2 < 0$. In this case there will be combinations of values of h and k for which we can get Q positive or negative. So in this case we have a *saddle point*.

Finally, if $D = f_{xx}f_{yy} - f_{xy}^2 = 0$, then we can say nothing about the nature of the function at the critical point – we need to go to higher-order terms in the Taylor series, and use another test.

So, to summarize, we have proved the following:

● *Second derivative test for local extreme values* ────────

Let $f(x, y)$ and its first and second partial derivatives be continuous in a neighbourhood of (x_0, y_0) and let $f_x(x_0, y_0) = f_y(x_0, y_0) = 0$. Then:

(i) f has a local maximum at (x_0, y_0) if

$$f_{xx} < 0 \text{ and } f_{xx}f_{yy} - f_{xy}^2 > 0$$

at (x_0, y_0);

(ii) f has a local minimum at (x_0, y_0) if

$$f_{xx} > 0 \text{ and } f_{xx}f_{yy} - f_{xy}^2 > 0$$

at (x_0, y_0);

(iii) f has a saddle point at (x_0, y_0) if

$$f_{xx}f_{yy} - f_{xy}^2 < 0$$

at (x_0, y_0);

(iv) no conclusion is possible at (x_0, y_0) if

$$f_{xx}f_{yy} - f_{xy}^2 = 0$$

at (x_0, y_0) – a further test is then needed.

Now, I remember that when I first saw this topic, I felt very insecure about all the second derivatives and the form of the discriminant. And I certainly did not see much of a pattern that could be extended to functions of more variables. But actually there is a pattern. Consider the array of second derivatives (remember $f_{xy} = f_{yx}$):

$$\begin{bmatrix} f_{xx} & f_{xy} \\ f_{xy} & f_{yy} \end{bmatrix}$$

Then notice that the conditions we have just derived depend on the sign of f_{xx} and

$$\begin{vmatrix} f_{xx} & f_{xy} \\ f_{xy} & f_{yy} \end{vmatrix}$$

the determinant of the array. This is the key. It is a classic result of linear algebra that the sign of the quadratic form in the Taylor series is dependent on such determinant conditions. This is the way in which the idea is generalized to functions of more than two variables, and although things do become more complicated, the determinant notation greatly simplifies things.

PROBLEM 9

> Classify all of the critical points determined in Problem 7.

(i) $f(x, y) = 6x - 3x^2 - y^2$

$$f_{xx} = -6 \qquad f_{yy} = -2 \qquad f_{xy} = 0$$

There is one critical point, at which we have for the discriminant,

$$D = f_{xx}f_{yy} - f_{xy}^2 = 12 > 0$$

Since $f_{xx} = -6 < 0$, then this is a local maximum. We can see this more directly by expanding about the critical point, by putting $x = 1 + h$, $y = k$, to obtain $f(1 + h, k) = 6(h + 1) - 3(h + 1)^2 - k^2 = 3 - 3h^2 - k^2 = f(1, 0) - h^2 - k^2$. Clearly, for any small values of h and k, the maximum of this occurs at $f(1, 0)$ and its value for non-zero h, k will always be less than this value.

(ii) $f(x, y) = x^2 + xy + 3x + 2y + 5$

$$f_{xx} = 2 \qquad f_{yy} = 0 \qquad f_{xy} = 1$$

So

$$D = -1 < 0$$

and therefore the critical point $(-2, 1)$ is a saddle point.

EXERCISES ON 6.4

1. Prove the first derivative test for the case when $f(x, y)$ has a local minimum.
2. Find the stationary points and the local extreme values for the following functions:

   (i) $f(x, y) = x^3 - 6xy + y^3$ 　　　　(ii) $f(x, y) = \dfrac{1}{y} - \dfrac{1}{x} - 4x + y$

   (ii) $f(x, y) = e^{-(x^2 + y^2 + 2x)}$ 　　　(iv) $f(x, y) = \dfrac{1}{x^2 + y^2 - 1}$

3. Discuss the stationary points of $f(x, y) = x^2 + kxy + y^2$ for different values of the constant k.

6.5 Maxima and minima with constraints: Lagrange multipliers

So far, we have emphasized the analogy of functions of two or more variables with the case of functions of single variables, and this has worked quite well in the case of turning values. However, in the case of two or more variables there are possibilities which have no counterpart in the single-variable case. One of the classic problems of mathematics is such a possibility – *constrained optimization*: that is, finding the critical points of a multi-variable function when its variables are constrained by one or more equations.

Such problems often arise in very natural geometrical contexts – such as determining the maximum volume of a rectangular box given that the total length of the edges is fixed.

PROBLEM 10

> Show that this geometrical problem is equivalent to the constrained optimization problem:
>
> maximize $f(x, y, z) = xyz$
> subject to $c(x, y, z) = 4(x + y + z) - 12\ell = 0$ where ℓ is a constant.

If x, y, z are the sides of the box, then the volume to be maximized is $f(x, y, z) = xyz$. The total perimeter is $4(x + y + z)$, and this must be fixed, giving a constraint of the form $4(x + y + z) = 12\ell$ say.

There is an 'easy' way to solve this problem which is to eliminate one variable, say z, from the function f using the constraint equation and minimize the resulting two-variable form for f.

PROBLEM 11

> Solve Problem 10 the 'easy' way.

We have, from the constraint c:

$$z = 3\ell - (x + y)$$

So

$$f(x, y, z) \equiv xy(3\ell - x - y) \equiv h(x, y)$$

say. Note that we also have the conditions x, y, $z > 0$.
The critical points are at the solutions of

$$h_x = h_y = 0$$

i.e.

$$3\ell y - 2xy - y^2 = 0$$
$$3\ell x - x^2 - 2xy = 0$$

Since x, y can be assumed to be non-zero, these give

$$3\ell - 2x - y = 0$$
$$3\ell - x - 2y = 0$$

which yield the (expected?!) solution $x = y = \ell$, and hence $z = \ell$. The box is therefore a cube, with volume ℓ^3.
In this case there is little need to look at the second derivatives – the physical

conditions of the problem persuade us that the result is a maximum, although there is no harm in checking!

The above method is not always feasible, because, for one thing, it may not be possible to solve the constraint equation in practice. Lagrange developed a more direct method which does not require this.

Consider the problem of finding the maxima or minima of the function

$$z = f(x, y)$$

subject to the constraint

$$c(x, y) = 0$$

Suppose the second equation defines y in terms of x, so that z becomes a function of x alone. Then z has a critical point if and only if

$$\frac{\partial z}{\partial x} = 0$$

or

$$\frac{\partial f}{\partial x} + \frac{\partial f}{\partial y}\frac{\partial y}{\partial x} = 0 \tag{i}$$

Also, substituting for y in the constraint equation will yield the *identity*

$$c(x, y(x)) \equiv 0$$

Differentiating with respect to x gives

$$\frac{\partial c}{\partial x} + \frac{\partial c}{\partial y}\frac{\partial y}{\partial x} = 0 \tag{ii}$$

The values of x, y at a critical value satisfy (i) and (ii) and hence also satisfy the equation

$$\frac{\partial f}{\partial x} + \lambda\frac{\partial c}{\partial x} + \left(\frac{\partial f}{\partial y} + \lambda\frac{\partial c}{\partial y}\right)\frac{\partial y}{\partial x} = 0$$

where λ is an arbitrary *function*. Now choose λ to satisfy

$$\frac{\partial f}{\partial y} + \lambda\frac{\partial c}{\partial y} \equiv \frac{\partial}{\partial y}(f + \lambda c) = 0 \tag{iii}$$

Then x, y must satisfy

$$\frac{\partial f}{\partial x} + \lambda\frac{\partial c}{\partial x} \equiv \frac{\partial}{\partial x}(f + \lambda c) = 0 \tag{iv}$$

The constraint equation and the equations (iii), (iv) thus provide three equations for the three variables x, y, λ (λ will not actually be needed). So the *method of Lagrange multipliers* may be expressed as follows:

To find the critical points of $f(x, y)$ subject to $c(x, y) = 0$:

(i) Consider the function

$$F(x, y) = f(x, y) + \lambda c(x, y)$$

where λ is an, as yet unknown, number, called the *Lagrange multiplier*.

(ii) Solve the following three equations for x, y and λ:

$$\frac{\partial F}{\partial x} = 0 \qquad \frac{\partial F}{\partial y} = 0 \qquad c(x, y) = 0$$

The extension of this approach to functions of more than two variables is slightly complicated by the fact that we can have more than one constraint. However, Problem 12 gets you thinking about this, following which I will give the general result.

PROBLEM 12

> Repeat Problem 11 using Lagrange multipliers.

This problem deals with functions of three variables, and has one constraint:

maximize $\quad f(x, y, z) = xyz$

subject to $\quad c(x, y, z) = 4(x + y + z) - 12\ell = 0.$

Following the natural extension of the Lagrange method described above, we introduce a multiplier λ and consider the function

$$F(x, y, z) = xyz + 4\lambda(x + y + z) - 12\lambda\ell$$

The optimization problem then reduces to solving the equations

$$F_x = yz + 4\lambda = 0 \qquad \text{(i)}$$

$$F_y = xz + 4\lambda = 0 \qquad \text{(ii)}$$

$$F_z = xy + 4\lambda = 0 \qquad \text{(iii)}$$

$$c = 4(x + y + z) - 12\ell = 0 \qquad \text{(iv)}$$

Subtracting equations (i)–(iii) in pairs and remembering that x, y, z must all be positive and therefore non-zero, we find

$$x = y = z$$

Substitution in (iv) then gives the common value to be ℓ, as before, giving a volume of ℓ^3. Notice that there was never any need to actually evaluate λ. Also, notice that the calculations were somewhat 'cleaner' than the previous elimination approach – effectively we have benefited from the fact that the Lagrange method maintained the symmetry in the variables x, y, z. Another advantage of the Lagrange method is that it yields all critical points, whereas some may be lost in the elimination method (cf. Jones and Jordan, 1970). Now, the promised generalization of Lagrange's method:

● *Lagrange's method of undetermined multipliers for the general case*

To find maxima and minima of the function

$$z = f(x_1, x_2, \ldots, x_n)$$

of the n variables $x_1, x_2, \ldots, x_n$ subject to the m constraints $(m < n)$,

$$c_1(x_1, x_2, \ldots, x_n) = 0$$
$$c_2(x_1, x_2, \ldots, x_n) = 0$$
$$\vdots$$
$$c_m(x_1, x_2, \ldots, x_n) = 0$$

construct the function

$$F(x_1, x_2, \ldots, x_n) = f + \sum_{j=1}^{m} \lambda_j c_j$$

where $\lambda_j, j = 1, \ldots, m$, are Lagrange multipliers and solve (if possible!) the $m + n$ equations

$$\frac{\partial F}{\partial x_i} = 0 \qquad i = 1, \ldots, n$$

$$c_j = 0 \qquad j = 1, \ldots, m$$

Explicitly these take the form

$$\frac{\partial f}{\partial x_i} + \sum_{j=1}^{m} \lambda_j \frac{\partial c_j}{\partial x_i} = 0 \qquad i = 1, \ldots, n$$

$$c_j = 0 \qquad i = 1, \ldots, m$$

The approach does not distinguish between the types of stationary values. Also, it can break down and lead to difficulties if the above equations are singular in the sense that they do not yield consistent solutions for the variables concerned. But you will be well past needing this book if you ever have to deal with such eventualities!

EXERCISES ON 6.5

1. Find the shortest distance from the point (x_0, y_0) to a line $ax + by + c = 0$, using Lagrange multipliers.
2. Find the maxima and minima of f subject to the constraint set given:

   (i) $f(x, y) = xy,$ $\qquad\qquad\qquad$ $c(x, y) = x^2 + 2y^2 - 4 = 0$

   (ii) $f(x, y) = x^2 + y^2,$ $\qquad\qquad\quad$ $c(x, y) = xy - 1 = 0$

   (iii) $f(x, y, z) = xy^2z^3,$ $\qquad\qquad$ $c(x, y, z) = x + y + z - 50 = 0$

   (iv) $f(x, y, z) = x^2 + 2y - z^2,$ $\qquad$ $c_1(x, y, z) = 2x - y = 0,$
   $\qquad\qquad\qquad\qquad\qquad\qquad\quad$ $c_2(x, y, z) = y + z = 0$

FURTHER EXERCISES

1. Expand the following functions in Maclaurin's series, up to terms of third order

   (i) xe^y $\qquad$ (ii) $x \cos y$ $\qquad$ (iii) $e^y \sin x$ $\qquad$ (iv) $e^{x^2 + y^2}$

   (v) $\ln(2x + y)$ $\qquad$ (vi) $\dfrac{1}{1 + x + y}$ $\qquad$ (vii) $\dfrac{x}{1 + xy}$

2. A function $f(x, y)$ is said to be *homogeneous of degree n* if it satisfies

$$f(\lambda x, \lambda y) = \lambda^n f(x, y)$$

By expanding both sides of

$$f(x + xt, y + yt) = (1 + t)^n f(x, y)$$

in powers of t deduce Euler's theorems for a homogeneous function:

$$xf_x + yf_y = nf$$
$$x^2 f_{xx} + 2xyf_{xy} + y^2 f_{yy} = n(n-1)f$$
$$x^3 f_{xxx} + 3x^2 yf_{xxy} + 3xy^2 f_{xyy} + y^3 f_{yyy} = n(n-1)(n-2)f$$

3. Find and classify the critical points of the following functions:
   (i) $f(x, y) = x^2 + y^2 + 2xy + x - y + 3$
   (ii) $f(x, y) = x^2 + y^2 - xy + 2x + 2y - 4$
   (iii) $f(x, y) = 3x^2 - 2xy + y^2 - 8y$
   (iv) $f(x, y) = x^2 + y^2 + xy + x + y + 1$
   (v) $f(x, y) = 2x^2 - 3y^2 + 2xy + x + 1$
   (vi) $f(x, y) = x^2 + 2y^2 - x^2 y$
   (vii) $f(x, y) = x^2 + y^2 - 2x - 4y - 5$
   (viii) $f(x, y) = 4x + 2y - x^2 + xy - y^2$

4. Locate and classify the critical points of the functions
   (i) $f(x, y) = (x + y)(xy + 1)$
   (ii) $f(x, y) = x^2 + y^2 + (ax + by + c)^2$
   (iii) $f(x, y) = x^3 + y^3 - 3xy$
   (iv) $f(x, y) = xy + \dfrac{2}{x} + \dfrac{4}{y}$
   (v) $f(x, y) = y \sin x$
   (vi) $f(x, y) = x^3 - x^2 y + y^2 - 2y + 1$

5. A rectangle is to be inscribed in the ellipse $x^2/a^2 + y^2/b^2 = 1$, with its sides parallel to the axes. Find the dimensions of such a rectangle if it is to have
   (i) the greatest area;
   (ii) the greatest perimeter.

6. Show that the triangle of largest perimeter which can be inscribed in a circle is equilateral.

7. Solve the following constrained optimization problems.
   (i) Maximize $f(x, y) = x^2 + y^2$, subject to $x^4 + 7x^2 y^2 + y^4 - 1 = 0$.
   (ii) Find the maximum and minimum values of $f(x, y) = 3x + 4y$, given that $x^2 + y^2 - 1 = 0$.
   (iii) Find the maximum and minimum values of $f(x, y) = 4x^3 + y^3$, subject to $2x^2 + y^2 - 1 = 0$.
   (iv) Minimize $f(x, y) = x + y$, subject to $xy - 16 = 0$, $x > 0$, $y > 0$.
   (v) Maximize $f(x, y, z) = 2x + 3y + 5z$, subject to $x^2 + y^2 + z^2 = 19$.
   (vi) Find the maximum and minimum values of $f(x, y, z) = xyz$, subject to $x^2 + y^2 + z^2 = 1$.
   (vii) Maximize $f(x, y, z) = xyz$, subject to $x + y + z = 30$, and $x + y - z = 0$.

8. Find the maximum value of $x_1 + x_2 + \ldots + x_n$ subject to $x_1^2 + x_2^2 + \ldots + x_n^2 = 1$.

9. Essay or discussion topics:
   (i) Extending the ideas and methods of convergence testing to Taylor series for functions of two variables.
   (ii) Stationary points beyond the quadratic term.
   (iii) Extreme values, critical points and their nature.
   (iv) The geometric significance of the Lagrange multiplier.

7 • Multiple Integration

7.1   Introduction

Just as we can have differentiation of functions of more than one variable, so we can also have integration of such functions. And as partial differentiation is essentially just differentiation with respect to one variable at a time, so 'multiple integration' is basically integration with respect to one variable at a time. However, as you will remember from calculus of a single variable, integration is always just that bit more awkward than differentiation – and so it is with multiple integration. The main problem is the treatment of the 'definite integral'. In ordinary integration the definite integral is an integral over some interval, $a \leq x \leq b$, of the real line. The only difficulty in evaluating such an integral resides in the nature of the integrand. In the case of functions of two variables this becomes an integration over a *region* in xy-space. In this case both the integrand and the region of integration can give problems in multiple integration. First, let us look at some basic elementary principles.

PROBLEM I

> Taking a common-sense view, what do you think we should do with
>
> (i) $\int x^2 y \, dx$ (ii) $\int x^2 y \, dy?$

Clearly, we could regard $\int x^2 y \, dx$ as the integral of $x^2 y$, with respect to x, with y held constant:

$$\int x^2 y \, dx = \frac{x^3 y}{3} + f(y)$$

Note that we have to add an arbitrary *function* of y, $f(y)$. In ordinary integration this would have been an arbitrary constant, but here we are regarding y as constant. Similarly,

$$\int x^2 y \, dy = \frac{x^2 y^2}{2} + g(x)$$

where $g(x)$ is an arbitrary function of x.

   This all seems very logical – but what happens when x and y are limited to take certain values on an interval? The simplest case is when x, y are defined on a rectangle $a \leq x \leq b$, $c \leq y \leq d$. In this case, what meaning would we give to, say, $\int_a^b x^2 y \, dx?$

It seems reasonable to take

$$\int_a^b x^2 y \, dx = \left[\frac{x^3}{3}y + f(y)\right]_a^b$$

$$= \frac{b^3}{3}y + f(y) - \frac{a^3}{3}y - f(y)$$

$$= \left(\frac{b^3 - a^3}{3}\right)y$$

The arbitrary function of y has cancelled out – just like the constant of integration in ordinary definite integration. Similarly, we would write

$$\int_c^d x^2 y \, dy = x^2\left(\frac{d^2 - c^2}{2}\right)$$

But all we are doing here is ordinary integration with respect to x (y), keeping y (x) constant. In the case of the above rectangular region we might venture further and define a *double integral* in the obvious way by

$$\int_a^b \int_c^d x^2 y \, dy \, dx$$

We can think of this as saying 'integrate with respect to y keeping x constant, then integrate the result with respect to x'.

PROBLEM 2

> Do the above integration.

$$\int_a^b \int_c^d x^2 y \, dy \, dx = \int_a^b \left[x^2\left(\frac{d^2 - c^2}{2}\right)\right] dx$$

$$= \left(\frac{b^3 - a^3}{3}\right)\left(\frac{d^2 - c^2}{2}\right)$$

This seems reasonable, but it already raises a number of questions for more complicated examples.

For example, what do we do if the region of integration is more complicated than a rectangle? And is it possible to 'reverse' the order of integration and do the x-integration first – and would the result be the same? The answer to the first question is what makes multiple integration difficult, the answer to the second is yes – sometimes, and with care. There are two standard tricks for dealing with integration (or other processes) over a region which is more complicated than a rectangle. One method is to break the region up into simpler regions such as small rectangles, do the process on the simpler regions and then combine the results. This is often feasible for the sorts of regions and the sorts of operations encountered in many areas of mathematics. Another method is to enclose the region on which the function is defined in a larger rectangular region, as described in Section 7.3. We can also have 'triple integrals' – or indeed 'higher-order' multiple integrals. In such cases, dealing with the region of integration is more difficult, of course, but the general principles are the same.

EXERCISE ON 7.1

Evaluate

$$\int_1^2 \int_0^1 (x+y)xy \, dx \, dy$$

7.2 Double integrals over a rectangle: volume under a surface

Here we are concerned with the integration of a function $f(x, y)$ defined on a rectangular region

$$R : a \le x \le b, \quad c \le y \le d \tag{7.1}$$

in rectangular coordinates. Note that the mathematical reason why this region is simple in rectangular coordinates is that the boundaries can be defined by adopting constant values for the coordinates. Were we using polar coordinates in the plane then a rectangle would be far from simple. Instead, a *disc* would be the simplest region, defined by $0 \le \theta < 2\pi, 0 \le r \le a$, say.

It is worth reminding ourselves how we deal with the definite integral of a function of a single variable:

$$\int_a^b f(x) \, dx$$

There are a number of ways of defining such an integral. One is to regard the integral as the reverse of differentiation and substitute the limits in the appropriate anti-derivative – if you can find it. This approach is equivalent to the methods which express the integral in terms of limits of sums. These latter methods do not rely on being able to evaluate the anti-derivative, and reflect the interpretation of the integral as an area. We can similarly define a 'double integral' in terms of limits of sums. Instead of dividing the interval $a \le x \le b$ into subintervals we divide the rectangle into cells by means of a series of lines parallel to the x- and y-axes (see Fig. 7.1).

In ordinary integration there are a number of approaches to expressing the integral as the limit of a sum. The interval, $a \le x \le b$, can be divided into regular or non-regular subintervals. And when evaluating the area under the curve we can take the strips on the subintervals to be rectangles or trapeziums or some other shape. Also, we can express the integral as a value sandwiched between a minimum and a maximum coverage of the area by strips protruding above the curve or resting

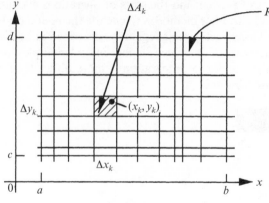

Fig. 7.1

below it. Each of these variations can be generalized to double integrals. Here we will use the latter approach of 'sandwiching' the value of the integral.

Let f be a function continuous on the rectangle R defined in (7.1). Now partition the intervals $[a, b]$, $[c, d]$ as follows:

$$P_1 = \{x_0, x_1, \ldots, x_m\} \text{ for } [a, b], \text{ with } x_0 = a, x_m = b$$
$$P_2 = \{y_0, y_1, \ldots, y_n\} \text{ for } [c, d], \text{ with } y_0 = c, y_n = d$$

Then the Cartesian product

$$P = P_1 \times P_2 = \{(x_i, y_j) : x_i \in P_1, y_j \in P_2\}$$

is a partition of R into $m \times n$ non-overlapping rectangles

$$R_{ij} : x_{i-1} \le x \le x_i, y_{j-1} \le y \le y_j$$

where $1 \le i \le m$, $1 \le j \le n$ (*see* Fig. 7.2).

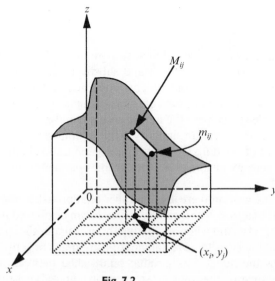

Fig. 7.2

On each rectangle R_{ij} the function f, being continuous, has a maximum value M_{ij} and a minimum value m_{ij}. Now form the *upper sum*

$$U_f(P) \equiv \sum_{i=1}^{m} \sum_{j=1}^{n} M_{ij}(x_i - x_{i-1})(y_j - y_{j-1})$$

$$\equiv \sum_{i=1}^{m} \sum_{j=1}^{n} M_{ij} \Delta x_i \Delta y_j$$

and the *lower sum*

$$L_f(P) = \sum_{i=1}^{m} \sum_{j=1}^{n} m_{ij} \Delta x_i \Delta y_j$$

We now define the *double integral* of f over R to be the unique number I that satisfies the inequality

$$L_f(P) \le I \le U_f(P)$$

for all partitions P of R. We denote this number by

$$I = \iint_R f(x, y)\,dx\,dy$$

Note that no particular order of integration is implied in this notation – we could as easily write $dy\,dx$. Indeed, we sometimes denote the integral by

$$I = \iint_R f(x, y)\,dA$$

where dA is an elemental area of the region R. We only need to consider the order of integration when we actually come to evaluate double integrals by treating them as iterated integrals.

If the above definition appears a little strange, first note that in the case when $z = f(x, y)$ stays positive over the region R, then each term $M_{ij} \Delta x_i \Delta y_j$ can be interpreted as the 'volume' of a rectangular column with base area $\Delta A_{ij} = \Delta x_i \Delta y_j$ and height M_{ij}. The sum $U_f(P)$ is thus an overestimate of the volume under the surface $z = f(x, y)$ and over the region R. Similarly, $L_f(P)$ is an underestimate of that volume. By considering every possible partition P, we range over every possible underestimate and every possible overestimate of the volume. Provided appropriate conditions are met, sandwiched in between these is a unique number I which is exactly equal to the volume – thus the double integral is equivalent to the volume under the surface $z = f(x, y)$ over the region R. This is directly analogous to interpreting the ordinary definite integral as an area under a curve. No one in their right mind would use the above definition to actually evaluate a double integral – we will see how to do this later. However, just to show that it does work and to illustrate the ideas involved, try the following problem.

PROBLEM 3

Use the above method to evaluate the integral

$$\iint_R 2\,dx\,dy$$

where R is the rectangle $a \le x \le b$, $c \le x \le d$.

In this case $f(x, y) = 2$ for all x, y and so $M_{ij} = m_{ij} = 2$ whatever the partition. So take arbitrary partitions

$$P_1 = \{x_0, x_1, \ldots, x_m\}$$
$$P_2 = \{y_0, y_1, \ldots, y_n\}$$

Then

$$U_f(P) = \sum_{i=1}^{m}\sum_{j=1}^{n} 2\Delta x_i\,\Delta y_j = 2\left(\sum_{i=1}^{m}\Delta x_i\right)\left(\sum_{j=1}^{n}\Delta y_j\right)$$

$$= 2(a - b)(c - d)$$

Similarly,

$$L_f(P) = 2(a - b)(c - d)$$

By definition

$$\iint_R 2\,dx\,dy = I$$

where

$$L_f(P) \le I \le U_f(P)$$

So

$$2(a - b)(c - d) \le I \le 2(a - b)(c - d)$$

or

$$I = \iint_R 2\,dx\,dy = 2(a - b)(c - d)$$

You can confirm that this is what you would obtain by evaluating

$$\int_c^d\int_a^b 2\,dx\,dy$$

in the 'obvious' way.

Of course, this method only works so simply in this case because of the simplicity of the integrand, but it does show that the definition makes sense. Also, in this case the m_{ij} and M_{ij} are easy to find – they are always 2. If the integrand is a function of x and y then finding m_{ij} and M_{ij} can be problematical and in this case it is easiest simply to choose a convenient point on the rectangle R_{ij} – say (x_i, y_i) – to use in the construction of the double sum.

Consider the volume under the plane $z = 2x + y$ and over the rectangle $0 \le x \le 2$, $1 \le y \le 2$. Partition each of these intervals into n subintervals of equal length and obtain an approximation for the volume as a double sum. You will need the results

$$\sum_{i=1}^{n} \sum_{j=1}^{n} 1 = n^2 \qquad \sum_{i=1}^{n} i = \frac{1}{2} n(n + 1)$$

Check your results by direct double integration.

7.3 Double integrals over general regions and other properties

Now consider the double integral of a function $f(x, y)$ defined over a region Ω in the xy-plane bounded by a finite number of continuous arcs $y = \phi_i(x)$, $x = \psi_j(y)$ (an arc is a parametrization of a curve). Such a region is called a *basic region*. We can always enclose such a region in a rectangular region R. We then define a new function $F(x, y)$ by

$$F(x, y) = \begin{cases} f(x, y) & \text{for } (x, y) \text{ in } \Omega \\ 0 & \text{for } (x, y) \text{ in } R \text{ but not in } \Omega \end{cases}$$

as illustrated in Fig. 7.3.

We now define the double integral of $f(x, y)$ over Ω by the integral

$$\iint_R F(x, y)\, dx\, dy \equiv \iint_\Omega f(x, y)\, dx\, dy$$

You may object that the integral on the left-hand side is not guaranteed to exist, because when we discussed the double integral over the rectangle we assumed that the integrand was continuous. Now $F(x, y)$ is *not* continuous in general – there may be discontinuities at the boundary of Ω. However, you would not be surprised to

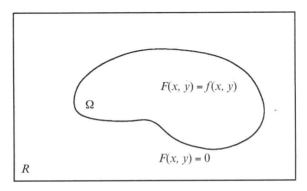

Fig. 7.3

learn that the arguments can be extended to cover the existence of integrals in such circumstances. In this case we say that $f(x, y)$ is *integrable* over Ω. Geometrically, we can still interpret the double integral

$$\iint_{\Omega} f(x, y) \, dx \, dy$$

as the volume 'below' $f(x, y)$ over the region Ω, provided $f(x, y)$ does not change sign over Ω.

While the above definition of the double integral over a non-rectangular region is fairly simple, it is hardly very practical, nor is it necessary. We can in fact use exactly the same method that we used for defining the integral over a rectangular region – partitioning the region up into small rectangles (or other suitable elemental areas adapted to the coordinate system). Thus, suppose the element has area ΔA_i, and (x_i^*, y_i^*) is some point within the area; then we can define the integral over the region as

$$I = \lim_{n \to \infty} \sum_{i=1}^{n} f(x_i^*, y_i^*) \Delta A_i$$

where n is the number of elements. Or, we can do as we did in Section 7.2 and sandwich I between over- and underestimates.

Note that if we take the function $z = f(x, y) = 1$ over the region Ω, then the volume under this surface is numerically equal to the area of Ω, so that in terms of a double integral, the *area* of Ω is given by

$$A = \iint_{\Omega} dx \, dy$$

In practice the definition of the integral over a general region, given above, is not a lot of use in actually evaluating the double integral. Indeed, all we have done so far is just *define* the double integral over a region. To actually evaluate such integrals, we expect to have to do a bit of integration. First, however, let us look at some properties of double integrals.

The fact that double integrals are defined in terms of double summations immediately leads us to:

● The linearity of double integrals ——————————————

If a and b are constants, and f and g are integrable functions defined over a region Ω, then

$$\iint_{\Omega} (af(x, y) + bg(x, y)) \, dx \, dy = a \iint_{\Omega} f(x, y) \, dx \, dy + b \iint_{\Omega} g(x, y) \, dx \, dy$$

This is exactly analogous to ordinary integration.

● Order ————————————————————————————

Another property that is easy to appreciate from the summation definition is that of *order* in double integrals. We would guess from the volume interpretation of the double integral that if $f(x, y)$ exceeds $g(x, y)$ at every point of the region Ω then the same would be true of the corresponding volumes under the surfaces. By noting that if each term of a summation exceeds the corresponding term in another summation then the total of the first exceeds the total of the second, we can prove that:

- if $f \geq 0$ on Ω, then

$$\iint_{\Omega} f(x, y) \, dx \, dy \geq 0$$

- if $f \geq g$ on Ω, then

$$\iint_{\Omega} f(x, y) \, dx \, dy \geq \iint_{\Omega} g(x, y) \, dx \, dy$$

● Additivity

A very useful result in ordinary integration relates to the additivity of definite integrals of the same function over different intervals. For example, if $a \leq b \leq c$, then we can write

$$\int_{d}^{c} f(x) \, dx = \int_{d}^{b} f(x) \, dx + \int_{b}^{c} f(x) \, dx$$

There is an analogous result for the double integral. If Ω is decomposed into two non-overlapping basic regions Ω_1, Ω_2, then, writing $\Omega = \Omega_1 + \Omega_2$ to mean the union of the two regions,

$$\iint_{\Omega} f(x, y) \, dx \, dy = \iint_{\Omega_1 + \Omega_2} f(x, y) \, dx \, dy = \iint_{\Omega_1} f(x, y) \, dx \, dy + \iint_{\Omega_2} f(x, y) \, dx \, dy$$

This is not difficult to appreciate from the summation definition – it is equivalent to combining all the 'cells' in region Ω_1 in a separate term of the summation and all the 'cells' in Ω_2 in another term. Clearly, the result can be extended to any number of component regions:

$$\iint_{\Omega_1 + \Omega_2 + \ldots + \Omega_n} f(x, y) \, dx \, dy = \iint_{\Omega_1} f(x, y) \, dx \, dy + \ldots + \iint_{\Omega_n} f(x, y) \, dx \, dy$$

One last property which is of theoretical importance is the following.

● Mean value theorem for double integrals

If f and g are continuous functions on a basic region Ω, and g is non-negative on Ω, then there exists a point (x_0, y_0) in Ω for which

$$\iint_{\Omega} f(x, y) \, g(x, y) \, dx \, dy = f(x_0, y_0) \iint_{\Omega} g(x, y) \, dx \, dy$$

PROOF

Being continuous on a closed and bounded region Ω, f must take on a minimum, m, and maximum, M, value on Ω. Since g is non-negative on Ω it then follows that

$$mg(x, y) \leq f(x, y) g(x, y) \leq Mg(x, y)$$

From the order result given above it then follows that

$$\iint_\Omega mg(x,y) \, dx \, dy \leq \iint_\Omega f(x, y) g(x, y) \, dx \, dy \leq \iint_\Omega Mg(x, y) \, dx \, dy$$

or

$$m \iint_\Omega g(x,y) \, dx \, dy \leq \iint_\Omega f(x, y) g(x, y) \, dx \, dy \leq M \iint_\Omega g(x, y) \, dx \, dy$$

Again, the order property tells us that

$$\iint_\Omega g(x, y) \, dx \, dy \geq 0$$

so we can divide through by this quantity without fear of reversing the inequalities. We can also exclude the equality, since the theorem clearly holds in that case, for all (x_0, y_0) in Ω. Otherwise, we can write

$$m \leq \frac{\iint_\Omega f(x, y)g(x, y) \, dx \, dy}{\iint_\Omega g(x, y) \, dx \, dy} \leq M$$

By an obvious extension of the *intermediate value theorem* for functions of a single variable, since m and M are minimum and maximum values of $f(x, y)$ respectively, it follows that there exists (x_0, y_0) in Ω for which

$$\frac{\iint_\Omega f(x, y)g(x, y) \, dx \, dy}{\iint_\Omega g(x, y) \, dx \, dy} = f(x_0, y_0)$$

which proves the theorem.

From the mean value theorem we can deduce the following result, which is very useful for estimating the value of a double integral. Let f be integrable over Ω. Suppose that f is bounded above and below by constants m, M respectively:

$$m \leq f(x, y) \leq M \qquad \forall (x, y) \in \Omega$$

Then

$$mA_\Omega \leq \iint_\Omega f(x, y) \, dx \, dy \leq MA_\Omega$$

where A_Ω is the area of Ω.

1. Let Ω be the disc $\{(x, y) : x^2 + y^2 \le 1\}$. Find upper and lower bounds for

(i) $\displaystyle\iint_\Omega \frac{1}{2 + 3(x^2 + y^2)} dx\, dy$ (ii) $\displaystyle\iint_\Omega e^{-(x^2 + y^2)} dx\, dy$

2. If Ω is the rectangle $\{(x, y) : 0 \le x \le 1, 1 \le y \le 2\}$, obtain lower and upper bounds for

$$\iint_\Omega (x^3 y + 3x^2 y^2)\, dx\, dy$$

7.4 The evaluation of double integrals: repeated integration

We now turn to the actual evaluation of double integrals. Of course, we would expect to have to integrate the integrand in some way – first with respect to one variable then with respect to the other. But in double integration we have to contend with the nature of the region over which we integrate, as well as the integrand itself. It turns out that such integrations are straightforward in principle if the region has one of two types of shape – either two edges are of the form $x = a$, $x = b$ (type I) or two edges are of the form $y = c$, $y = d$ (type II) – *see* Fig. 7.4.

The point about such regions is that at least one of the integrals in the double integral has constant limits. Many regions occurring in practice are of one of these types, and those that are not can be split into parts which are and to these we can apply the additivity result of the previous section. To avoid complications, we will assume that $z \ge 0$ so that the surface does not dip below the xy-plane.

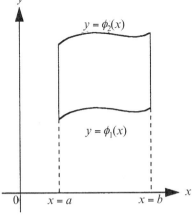

Type I region

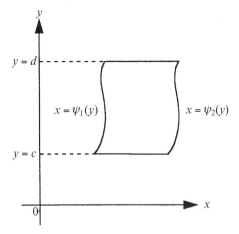

Type II region

Fig. 7.4

● Double integration over a type I region ————————————

In this case Ω consists of all points satisfying

$$a \le x \le b, \qquad \phi_1(x) \le y \le \phi_2(x)$$

The task is to evaluate the double integral

$$\iint_\Omega f(x, y)\, dx\, dy$$

by integration. The interpretation of the double integral as a volume provides a clue as to how this can be done. The volume could be evaluated by considering slices of the solid contained between $z = f(x, y)$ and Ω parallel to the yz-plane – i.e. by taking fixed values of x. We evaluate the area of each such slice, and then combine these by integrating over x. The area of the slice at x is

$$A(x) = \int_{\phi_1(x)}^{\phi_2(x)} f(x, y)\, dy \tag{7.2}$$

It is by no means obvious, but it can be proved that the total volume is then

$$\iint_\Omega f(x, y)\, dx\, dy = \int_a^b A(x)\, dx = \int_a^b \left(\int_{\phi_1(x)}^{\phi_2(x)} f(x, y)\, dy \right) dx$$

which gives a means of evaluating the double integral over a type I region by a *repeated* or *iterated integral*. We first integrate $f(x, y)$ with respect to y to form equation (7.2) and then integrate this with respect to x from a to b. We can deal with a type II region in a similar way.

● Double integration over a type II region ————————————

In this case Ω is defined by

$$c \le y \le d, \qquad \psi_1(y) \le x \le \psi_2(y)$$

and, in an obvious way,

$$\iint_\Omega f(x, y)\, dx\, dy = \int_c^d \left(\int_{\psi_1(y)}^{\psi_2(y)} f(x, y)\, dx \right) dy$$

In this case we integrate first with respect to x and then integrate the result with respect to y.

Note that it is usual to omit the inner brackets and write

$$\int_a^b \left(\int_{\phi_1(x)}^{\phi_2(x)} f(x, y)\, dy \right) dx \equiv \int_a^b \int_{\phi_1(x)}^{\phi_2(x)} f(x, y)\, dy\, dx$$

$$\int_c^d \left(\int_{\psi_1(y)}^{\psi_2(y)} f(x, y)\, dx \right) dy \equiv \int_c^d \int_{\psi_1(y)}^{\psi_2(y)} f(x, y)\, dx\, dy$$

In applying these results in practice, the first hurdle is to sort out your ideas on the region Ω. Those of you who avoided the difficulties of curve sketching in your elementary mathematics may now have cause to regret it! *Always* sketch the region before doing the integration.

Evaluate the integral

$$\iint_{\Omega} dx\, dy$$

where Ω is the region $\{(x, y) : 1 \le x \le 3, 2 \le y \le 4\}$.

In this case $f(x, y) = 1$ and the region Ω is the rectangle shown in Fig. 7.5. This is of both type I and type II, and we could evaluate the repeated integrals in either order. We will look at this later in Section 7.6, but for now let us just focus on the evaluation of the integral.

Treating it as a type I integral, we have

$$\iint_{\Omega} dx\, dy = \int_1^3 \left(\int_2^4 dy \right) dx = \int_1^3 2\, dx = 2(3 - 1) = 4$$

Treating it as a type II integral, we have

$$\iint_{\Omega} dx\, dy = \int_2^4 \left(\int_1^3 dx \right) dy = \int_2^4 2\, dy = 4$$

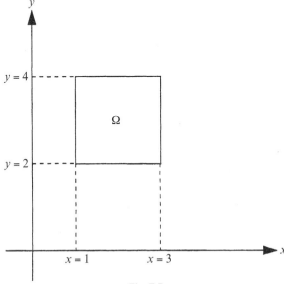

Fig. 7.5

Notice that the results are what we would expect from the volume interpretation of the integral. The result should be the volume of a solid of uniform height 1 erected on a base region of area $2 \times 2 = 4$ – i.e. 4 cubic units. Note that this also gives us the numerical value of the area of the region Ω, as it should.

You might also notice that we could equally well have written, in this case,

$$\iint_{\Omega} \mathrm{d}x\,\mathrm{d}y = \left(\int_1^3 \mathrm{d}x\right)\left(\int_2^4 \mathrm{d}y\right) = 2 \times 2 = 4$$

This result can be generalized for integrals of separated variables over a rectangle. Thus if R is the rectangle $a \le x \le b$, $c \le y \le d$ and f, g are continuous on $[a, b]$, $[c, d]$ respectively, then

$$\int_c^d \int_a^b f(x)g(y)\,\mathrm{d}x\,\mathrm{d}y = \int_c^d \left(\int_a^b f(x)g(y)\,\mathrm{d}x\right)\mathrm{d}y$$

$$= \int_c^d g(y)\left(\int_a^b f(x)\,\mathrm{d}x\right)\mathrm{d}y = \left(\int_a^b f(x)\,\mathrm{d}x\right)\left(\int_c^d g(y)\,\mathrm{d}y\right)$$

That is, we can reduce the double integral to a product of ordinary integrals in such circumstances.

PROBLEM 5

Evaluate

$$\iint_{\Omega} (x + y)\,\mathrm{d}x\,\mathrm{d}y$$

where Ω is the region
(i) $\{(x, y) : 0 \le x \le 1,\ x^2 \le y \le 2x\}$
(ii) $\{(x, y) : 0 \le y \le 1,\ y^2 + 1 \le x \le y^2 + 2\}$.

(i) This is a type I region and is illustrated in Fig. 7.6. In terms of repeated integrals we have

$$\iint_{\Omega}(x+y)\,\mathrm{d}x\,\mathrm{d}y = \int_0^1\left(\int_{x^2}^{2x}(x+y)\,\mathrm{d}y\right)\mathrm{d}x = \int_0^1\left[xy + \frac{y^2}{2}\right]_{x^2}^{2x}\mathrm{d}x = \int_0^1\left(4x^2 - x^3 - \frac{x^4}{2}\right)\mathrm{d}x = \frac{23}{24}$$

(ii) This is a type II region, sketched in Fig. 7.7. In this case

$$\iint_{\Omega}(x+y)\,\mathrm{d}x\,\mathrm{d}y = \int_0^1\left(\int_{y^2+1}^{y^2+2}(x+y)\,\mathrm{d}x\right)\mathrm{d}y = \int_0^1\left[\frac{x^2}{2} + xy\right]_{y^2+1}^{y^2+2}\mathrm{d}y$$

$$= \int_0^1\left(y^2 + y + \frac{3}{2}\right)\mathrm{d}y = \frac{7}{3}$$

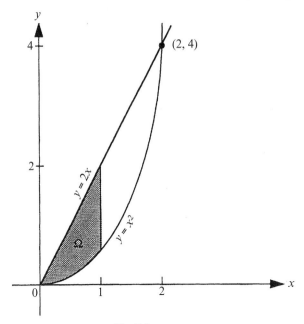

Fig. 7.6

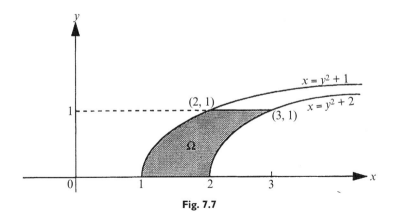

Fig. 7.7

In working through these problems, I suspect that the most difficult part for you will be in figuring out the limits on the integrals. This is in fact the most difficult part of multiple integration – it does not really give us any trouble in the case of functions of a single variable. But actually, provided you are systematic, it is not so bad for double integrals either.

Thus, to evaluate the double integral

$$\iint_\Omega f(x, y) \, dx \, dy$$

over a region Ω, integrating first with respect to y and then with respect to x, follow the procedure below, illustrated in Fig. 7.8 for the shaded area between the curves $x + y = 2$, $x^2 + y^2 = 4$.

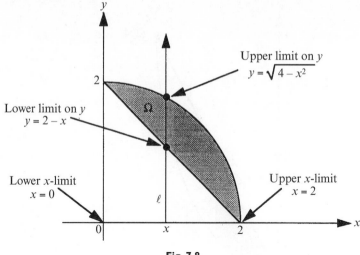

Fig. 7.8

(i) First sketch the region.

(ii) To determine the y-limits of integration consider a vertical line, ℓ, cutting the region, in the direction of increasing y. The values of y, for general x, where the line enters and leaves the region give the lower and upper limits of the y-integration respectively.

(iii) To determine the x-limits of integration we take the values of x which include all vertical lines through Ω.

Finally, we can write the integral as

$$\iint_{\Omega} f(x, y) \, dx \, dy = \int_{x=0}^{x=2} \int_{y=2-x}^{y=\sqrt{4-x^2}} f(x, y) \, dy \, dx$$

An alternative way to view this is that we divide the region of integration first into vertical strips, taking the y-integral over these between the lower and upper limits $y = 2 - x$ and $y = \sqrt{4 - x^2}$, to give

$$\int_{2-x}^{\sqrt{4-x^2}} f(x, y) \, dy$$

and then we integrate the results over all possible values of x to obtain

$$\int_{0}^{2} \int_{2-x}^{\sqrt{4-x^2}} f(x, y) \, dy \, dx$$

as before.

If we want to integrate in the reverse order, then we consider a horizontal line and look at the x-values where it enters and leaves the region.

PROBLEM 6

> Reverse the order for the above integral.

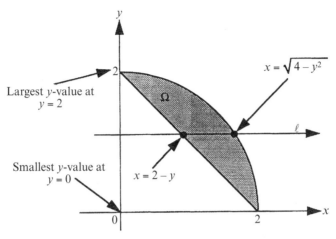

Fig. 7.9

As illustrated in Fig. 7.9, the horizontal line ℓ enters the region at $x = 2 - y$ and leaves it at $x = \sqrt{4 - y^2}$ (in increasing values of x). Meanwhile, y varies between $y = 0$ and $y = 2$, and we get the repeated (or iterated) integral in reverse order

$$\iint_\Omega f(x, y) \, dx \, dy = \int_0^2 \int_{2-y}^{\sqrt{4-y^2}} f(x, y) \, dx \, dy$$

In this example the region is of both type I and type II, and there is no particular difficulty in 'reversing the order of integration'. If we consider regions of one or other type then the principles for finding the limits are the same, but it is clearly easier to integrate in one order than another. Thus, for a type I region $\{(x, y): a \leq x \leq b, \phi_1(x) \leq y \leq \phi_2(x)\}$ we draw a vertical line through an arbitrary x in $[a, b]$ and the points where it cuts the region give the limits on the first y-integral and we then let x range over $[a, b]$ for the second integral. If you try to reverse this and draw a horizontal line across the region, then in general the upper and lower integration limits so defined are dependent in form on where you draw the line.

EXERCISES ON 7.4

1. Evaluate the double integrals

(i) $\displaystyle\int_0^1 \int_{x^2}^x xy^2 \, dy \, dx$ (ii) $\displaystyle\int_0^1 \int_{-\sqrt{1-y^2}}^{\sqrt{1-y^2}} x^2 y \, dx \, dy$

In each case sketch the region over which the integration is taken, and state what type it is.

2. Find the volume of the prism whose base is the triangle in the xy-plane bounded by the x-axis and the lines $y = x$ and $x = 1$ and whose top lies in the plane $z = 2 - x - y$.

3. Evaluate the following double integrals by means of an appropriate repeated integral:

   (i) $\iint_\Omega (x^2 + y^2)\,dx\,dy$, where $\Omega = \{(x, y) : 1 \le x \le 2, -1 \le y \le 1\}$

   (ii) $\iint_\Omega (x^2 + y)\,dx\,dy$, where Ω is the region in the first quadrant bounded by the x-axis, y-axis and the unit circle.

7.5 Reversing the order of integration

As noted in the previous section, when the region Ω is of both type I and type II, then we have the option of performing the repeated integral in either order – x first then y, or y first then x. This facility may be of real practical use in that it may be easier to use one order than the other. Here we will formalize the results illustrated in Section 7.4. Note that we are not really adding anything new here. The theorem I am about to state is the classic theorem which legitimizes the evaluation of multiple integrals by treating them as iterated integrals for regions which are either of type I or type II. We saw roughly how this comes about in Section 7.4. The facility to reverse the order of integration comes about simply from the region being both of type I and type II.

That we can reverse the order of integration in a repeated integral, under certain conditions, is guaranteed by:

● *Fubini's theorem* ─────────────────────────────────────

Let $f(x, y)$ be continuous on a region Ω. Then

(i) If Ω is a type I region defined by $a \le x \le b$, $\phi_1(x) \le y \le \phi_2(x)$, with ϕ_1 and ϕ_2 continuous on $[a, b]$, then

$$\iint_\Omega f(x, y)\,dx\,dy = \int_a^b \int_{\phi_1(x)}^{\phi_2(x)} f(x, y)\,dy\,dx$$

(ii) If Ω is a type II region defined by $c \le y \le d$, $\psi_1(y) \le x \le \psi_2(y)$, with ψ_1 and ψ_2 continuous on $[c, d]$, then

$$\iint_\Omega f(x, y)\,dx\,dy = \int_a^d \int_{\psi_1(y)}^{\psi_2(y)} f(x, y)\,dx\,dy$$

For regions that are of both type I and type II it then follows immediately that the iterated integral may be performed in any order, as noted above. Note the left-hand sides in the above equations – as noted in Section 7.2, the order of $dx\,dy$ has no significance here.

PROBLEM 7

Evaluate the integral

$$\int_0^1 \int_y^1 e^{x^2}\, dx\, dy$$

The point about this integral is that it is *impossible* to perform the x-integral first in terms of elementary functions. So let us see what happens if we reverse the order of integration. The region of integration is shown in Fig. 7.10.

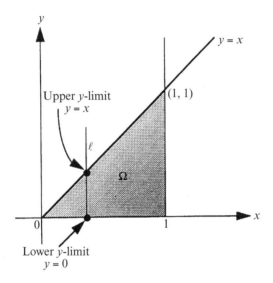

Fig. 7.10

To reverse the order of integration, treat the region as a type I region, with the y-integration ranging from $y = 0$ to $y = x$ and x ranging from 0 to 1 to give

$$\int_0^1 \int_y^1 e^{x^2}\, dx\, dy = \int\int_\Omega e^{x^2}\, dx\, dy = \int_0^1 \int_0^x e^{x^2}\, dy\, dx$$

$$= \int_0^1 \left[y e^{x^2} \right]_{y=0}^{y=x} dx = \int_0^1 x e^{x^2}\, dx = \frac{e - 1}{2}$$

This provides a very convincing illustration of the power of reversing the order of integration.

EXERCISES ON 7.5

1. In the following, reverse the order of integration:

(i) $\displaystyle\int_0^3\int_0^{\sqrt{x}} f(x, y)\, dy\, dx$

(ii) $\displaystyle\int_0^1\int_{-\sqrt{1-y^2}}^{\sqrt{1-y^2}} f(x, y)\, dx\, dy$

(iii) $\displaystyle\int_0^1\int_{-y}^{y} f(x, y)\, dx\, dy$

(iv) $\displaystyle\int_0^2\int_1^{e^x} f(x, y)\, dy\, dx$

(v) $\displaystyle\int_2^4\int_1^{y} f(x, y)\, dx\, dy$

2. Evaluate the following integrals using both orders of integration:

(i) $\displaystyle\int_0^1\int_2^{4-2x} dy\, dx$

(ii) $\displaystyle\int_1^2\int_0^{\ln y} e^{-x}\, dx\, dy$

(iii) $\displaystyle\int_2^4\int_1^{y}\frac{y^3}{x^3}\, dx\, dy$

3. If Ω is the triangle in the xy-plane bounded by the x-axis, the line $y = x$ and the line $x = 1$, evaluate the following integrals:

(i) $\displaystyle\iint_\Omega \frac{\sin x}{x}\, dx\, dy$

(ii) $\displaystyle\iint_\Omega \cos(\pi x^2/2)\, dx\, dy$

7.6 Double integrals in polar coordinates

When we viewed the double integral as a volume above a region in the xy-plane, you may have wondered if the way we divided up the region (into a grid of rectangles) is really that important – could we have divided it into triangles, for example? The answer is yes (indeed, triangles are the most useful partitioning areas, leading to what is called *triangulation* in topology), we could have, and so it is not that important – but it is much more *convenient* to use rectangles. This is because we are using rectangular coordinates to cover the plane. But what if we used polar coordinates (r, θ)? We know that this is often useful if we have circular symmetry in the plane. We could, of course, always convert from polar to rectangular coordinates and proceed as before, but this may make the integration difficult, and in fact it is not too difficult to handle double integrals in polar coordinates directly.

The trick is to find an *elemental area* in the plane, dA, which conforms nicely to polar coordinates, and replaces the rectangular grid in 'tiling' the plane. We obtained the rectangular grid by taking coordinate lines, $x = $ constant, $y = $ constant.

PROBLEM 8

> Obtain an expression for the elemental area dA in terms of polar coordi-
> nates by considering a small area bounded by coordinate lines between fixed
> coordinate values r, $r + dr$ and θ, $\theta + d\theta$.

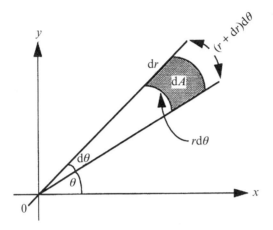

Fig. 7.11

Figure 7.11 illustrates the approach. The area required, dA, is illustrated. Its edges
have lengths $r\,d\theta$, dr, $(r + dr)\,d\theta$ and dr. Strictly of course, we should be using increments Δr, $\Delta\theta$, not differentials, but I do not want rigour to get in the way of the
method here, so we will use a little notational licence. There are a number of ways
of working out the area dA, but the easiest is to notice that it is essentially a
rectangle with sides dr, $r\,d\theta$ to first order in dr, dθ, giving an area

$$dA = r\,dr\,d\theta$$

If we now consider a function of polar coordinates, $f(r, \theta)$, then the volume of a
column above dA and under the surface $z = f(r, \theta)$ is just

$$dV = f(r, \theta)r\,dr\,d\theta$$

The total volume over a region Ω and under $z = f(r, \theta)$ is thus

$$V = \iint_\Omega f(r, \theta)r\,dr\,d\theta$$

The *area* of Ω is given by

$$A = \iint_\Omega r\,dr\,d\theta$$

By far the most common mistake that beginners make is to forget the extra r in the
double integral using polar coordinates.

Again, the most difficult part of doing a double integral in polar coordinates is to
find the integration limits. It is actually perhaps a little easier in polar coordinates

because there is really only one way to do it – we invariably have to evaluate the
r-integral first, and the θ-integral must be between constant limits. That is, the
usual region of integration in polar coordinates takes the form $\alpha \le \theta \le \beta, r_1(\theta) \le r \le$
$r_2(\theta)$, where α, β are constants. The procedure for finding the limits is similar to that
for rectangular coordinates (*see* Fig. 7.12):

  (i) Sketch the region.
 (ii) To get the r-limits consider a ray ℓ from the origin, at an angle θ, through the
      region. Then the lower and upper limits of the r-integration $r_1(\theta)$, $r_2(\theta)$ are the
      points where the ray enters and leaves the region respectively.
(iii) To get the θ-limits of integration take the smallest and largest values of θ that
      bound the region, say $\alpha \le \theta \le \beta$.

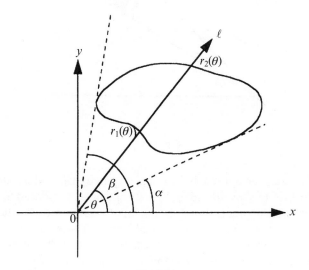

Fig. 7.12

Putting all this together gives

$$\iint_\Omega f(r, \theta) \, \mathrm{d}A = \int_\alpha^\beta \int_{r_1(\theta)}^{r_2(\theta)} f(r, \theta) r \, \mathrm{d}r \, \mathrm{d}\theta$$

From here on the evaluation of the double integral by repeated or iterated integration is exactly the same as for rectangular coordinates.

PROBLEM 9

Calculate the area enclosed by the cardioid

$r = 1 + \sin \theta$

shown in Fig. 7.13.

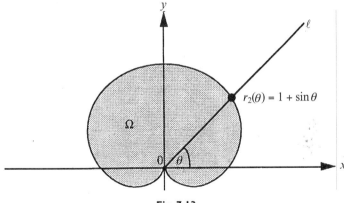

Fig. 7.13

The r-value ranges from 0 to $1 + \sin\theta$, while θ ranges from 0 to 2π, giving for the area the double integral:

$$A = \int\limits_0^{2\pi} \int\limits_0^{1+\sin\theta} r \, dr \, d\theta = \int\limits_0^{2\pi} \frac{1}{2}(1 + \sin\theta)^2 \, d\theta = \frac{3\pi}{2}$$

PROBLEM 10

Integrate the function $f(r, \theta) = r^2$ over the part of the unit disc, centre the origin, in the first quadrant (*see* Fig. 7.14).

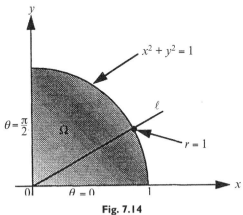

Fig. 7.14

The r-limits are 0, 1 and the θ-limits are 0, $\pi/2$, and so we have

$$\iint\limits_\Omega r^2 \, dA = \int\limits_0^{\pi/2} \int\limits_0^1 r^2 \cdot r \, dr \, d\theta = \int\limits_0^{\pi/2} \int\limits_0^1 r^3 \, dr \, d\theta = \int\limits_0^{\pi/2} \frac{1}{4} \, d\theta = \frac{\pi}{8}$$

● *Conversion from rectangular to polar coordinates* ─────────

Sometimes, we are given an integral in rectangular coordinates and wish to convert to polar coordinates. This can simplify the integral, particularly if functions of $x^2 + y^2$ occur, because this may be replaced by r^2, a single variable. We can do this transformation by replacing $x = r \cos \theta$ and $y = r \sin \theta$ in the double integral $\iint_\Omega f(x, y) \, dA$ and taking the elemental area to be $dA = r \, dr \, d\theta$ in polar coordinates. Of course, the integration limits also change. Thus we obtain

$$\iint_\Omega f(x, y) \, dx \, dy = \iint_\Omega f(r \cos \theta, r \sin \theta) \, r \, dr \, d\theta$$

This is nothing other than ordinary substitution in integration, analogous to the single-variable case.

PROBLEM 11

Evaluate

$$\iint_\Omega e^{x^2 + y^2} \, dx \, dy$$

over the semicircular region bounded by the x-axis and the curve $y = \sqrt{1 - x^2}$.

We cannot integrate this in rectangular coordinates – but it comes out surprisingly easily if we transform to polar coordinates. Putting $x = r \cos \theta$, $y = r \sin \theta$ and replacing $dx \, dy$ by $r \, dr \, d\theta$ gives

$$\iint_\Omega e^{x^2 + y^2} \, dx \, dy = \int_0^\pi \int_0^1 e^{r^2} r \, dr \, d\theta$$

$$= \int_0^\pi \frac{1}{2}(e - 1) \, d\theta = \frac{\pi}{2}(e - 1)$$

EXERCISES ON 7.6

1. Find the area enclosed by the lemniscate $r^2 = 4 \cos 2\theta$.
2. Find the volume enclosed by the sphere $x^2 + y^2 + z^2 = a^2$.
3. Evaluate

$$\iint_\Omega \sin \theta \, dA$$

   where Ω is the region in the first quadrant that is outside the circle $r = 1$ and inside the cardioid $r = 1 + \cos \theta$.
4. Evaluate the following by changing to polar coordinates:

   (i) $\displaystyle\int_0^1 \int_0^{\sqrt{1-x^2}} \sin(x^2 + y^2) \, dy \, dx$

   (ii) $\displaystyle\int_0^2 \int_0^{\sqrt{4-x^2}} \frac{dy \, dx}{(1 + x^2 + y^2)^{3/2}}$

7.7 Surface area

In this section we look at how to work out the area of a surface defined by a function $z = f(x, y)$. Specifically, we derive a result for the surface area S of a surface $z = f(x, y)$ above a given region, Ω, in the xy-plane. We will revisit this in Section 10.4, where we will give a range of methods for determining surface area, and also define general surface integrals. This introductory section is essentially an application of double integration. We use an approach which is common in dealing with curved surfaces – we divide the surface into small flat plates and find the area of each such plate. So the first step is to find the area of a plane surface, above a rectangular region $\Omega = R$ of sides ℓ, w in the xy-plane, with sides parallel to the axes. Let the plane be given by

$$z = ax + by + c$$

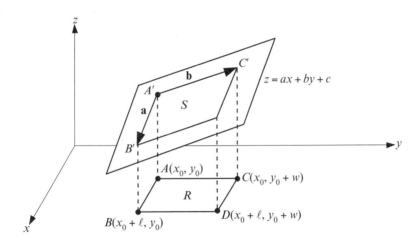

Fig. 7.15

Suppose the corners of the region R are

$$A(x_0, y_0), B(x_0 + \ell, y_0), C(x_0, y_0 + w), D(x_0 + \ell, y_0 + w)$$

The points A', B', C' corresponding to A, B, C on the plane surface above R are

$$A'(x_0, y_0, ax_0 + by_0 + c)$$

$$B'(x_0 + \ell, y_0, a(x_0 + \ell) + by_0 + c)$$

$$C'(x_0, y_0 + w, ax_0 + b(y_0 + w) + c)$$

These points determine a parallelogram with sides

$$\mathbf{a} = \overrightarrow{A'B'} = \ell\mathbf{i} + a\ell\mathbf{k} \qquad \mathbf{b} = \overrightarrow{A'C'} = w\mathbf{j} + bw\mathbf{k}$$

The area of the parallelogram is then

$$|\mathbf{a} \times \mathbf{b}| = |-a\ell w\mathbf{i} - \ell bw\mathbf{j} + \ell w\mathbf{k}| = \sqrt{a^2 + b^2 + 1}\ \ell w$$

Now consider an arbitrary region Ω, and divide this into small rectangles by taking lines parallel to the axes. Let R_k be a typical elemental rectangle in the xy-plane, with sides dx, dy (again, for full rigour these should be increments). Let ΔS_k be the area of the patch on the surface above R_k, and let (x, y) be any point in R_k. We can approximate ΔS_k by the area of the portion of the tangent plane at (x, y), above R_k. This tangent plane has equation

$$z = f_x x + f_y y + c$$

for some constant c (*see* Section 5.5), and so we have

$$\Delta S_k = \sqrt{f_x^2 + f_y^2 + 1}\ dx\,dy$$

The total surface area above Ω is then

$$S = \iint_\Omega \sqrt{f_x^2 + f_y^2 + 1}\ dx\,dy$$

PROBLEM 12

> Find the surface area of the circular paraboloid $z = x^2 + y^2$ between the xy-plane and the plane $z = 4$ shown in Fig. 2.1.

The surface area required is shown in Fig. 2.1. The region Ω is the disc $x^2 + y^2 \le 4$. We have

$$f_x = 2x \qquad f_y = 2y$$

so the required area is

$$S = \iint_\Omega \sqrt{1 + 4x^2 + 4y^2}\ dx\,dy$$

This is most easily evaluated using polar coordinates:

$$S = \int_0^{2\pi} \int_0^3 \sqrt{1 + 4r^2}\ r\,dr\,d\theta = \frac{\pi}{6}(17^{3/2} - 1)$$

As mentioned above, there are a number of approaches to surface areas and integrals, depending on the way in which the surface is represented. Here we have used the *explicit* representation $z = f(x, y)$. Other forms of representation are *implicit* and *parametric*, and we see how to deal with these in Section 10.4.

Find the surface area of the part of the parabolic cylinder $z = y^2$ that sits above the triangular region in the xy-plane enclosed by the x-axis, the line $y = x$ and the line $x = 1$.

7.8 Triple integrals

I have often referred to how results with two variables can be easily extended to more variables by tagging the extra variables on in an obvious way. So it is with the double integral, which may be formally extended to a *triple integral*,

$$\iiint_V f(x, y, z)\,\mathrm{d}x\,\mathrm{d}y\,\mathrm{d}z$$

where V is now a region in xyz-space (i.e. a 'volume'). More generally, the double integral

$$\iint_\Omega f(x, y)\,\mathrm{d}A$$

where $\mathrm{d}A$ is an element of area in the plane, extends naturally to the triple integral

$$\iiint_V f(x, y, z)\,\mathrm{d}\tau$$

where $\mathrm{d}\tau$ is an element of volume in 3-space. Just as $\mathrm{d}A$ differs for different coordinate systems, so $\mathrm{d}\tau$ will depend on the three-dimensional coordinate system used.

Since the analogies between double and triple (and indeed multiple) integrals are so natural, I will be very brief about developing the theory and methods of triple integrals. Just as we started with double integrals over a rectangle, so we will introduce triple integrals over a parallelepiped P in $\mathbb{R}^3$:

$$P = \{(x, y, z) : a_1 \leq x \leq a_2, b_1 \leq y \leq b_2, c_1 \leq y \leq c_2\}$$

We divide the intervals (a_1, a_2), (b_1, b_2), (c_1, c_2) into partitions of equal width:

$$a_1 = x_0 < x_1 < \ldots < x_n = a_2$$

$$b_1 = y_0 < y_1 < \ldots < y_m = b_2$$

$$c_1 = z_0 < z_1 < \ldots < z_p = c_2$$

This produces elemental volumes in the form of nmp boxes, each with typical volume

$$\Delta\tau = \Delta x\,\Delta y\,\Delta z$$

Let (x_i^*, y_j^*, z_k^*) be a typical point in a typical box and let

$$\Delta u = \sqrt{\Delta x^2 + \Delta y^2 + \Delta z^2}$$

Then if

$$\lim_{\Delta u \to 0} \sum_{i=1}^{n} \sum_{j=1}^{m} \sum_{k=1}^{p} f(x_i^*, y_j^*, z_k^*) \, \Delta x \, \Delta y \, \Delta z$$

exists and is independent of the way in which the points (x_i^*, y_j^*, z_k^*) are chosen, then we define the *triple integral* of f over P as

$$\iiint_P f(x, y, z) \, d\tau = \lim_{\Delta u \to 0} \sum_{i=1}^{n} \sum_{j=1}^{m} \sum_{k=1}^{p} f(x_i^*, y_j^*, z_k^*) \, \Delta \tau$$

As with double integrals, we can write this in the form of a repeated or iterated integral

$$\iiint_P f(x, y, z) \, d\tau = \int_{a_1}^{a_2} \int_{b_1}^{b_2} \int_{c_1}^{c_2} f(x, y, z) \, dz \, dy \, dx$$

PROBLEM 13

Evaluate $\iiint_P xze^{xy} \, d\tau$, where P is the parallelepiped $\{(x, y, z) : 0 \le x \le 1, 0 \le y \le 1, 0 \le z \le 2\}$.

$$\iiint_P xze^{xy} \, d\tau = \int_0^1 \int_0^1 \int_0^2 xze^{xy} \, dz \, dy \, dx = \int_0^1 \int_0^1 2xe^{xy} \, dy \, dx$$

$$= 2\int_0^1 \left[e^{xy} \right]_{y=0}^{y=1} \, dx = 2\int_0^1 (e^x - 1) \, dx = 2(e - 2)$$

The triple integral over a general region Π may be defined by the same trick as we used for double integrals. Thus, if we wish to define a triple integral over a bounded region Π, then we enclose Π in a parallelepiped P and define a new function F by

$$F(x, y, z) = \begin{cases} f(x, y, z) & \text{if } (x, y, z) \in \Pi \\ 0 & \text{if } (x, y, z) \notin \Pi \end{cases}$$

We then define

$$\iiint_\Pi f(x, y, z) \, d\tau = \iiint_P F(x, y, z) \, d\tau$$

The limits of integration of the corresponding iterated integral (whose existence is ensured by a three-dimensional form of Fubini's theorem) may be more difficult to find, but, referring to Fig. 7.16, we should be able to write down something of the form

$$\iiint_\Pi f(x, y, z) \, d\tau = \int_{x_1}^{x_2} \int_{y_1(x)}^{y_2(x)} \int_{z_1(x, y)}^{z_2(x, y)} f(x, y, z) \, dz \, dy \, dx$$

In this case we are 'projecting down' onto the xy-plane to do the integration. To

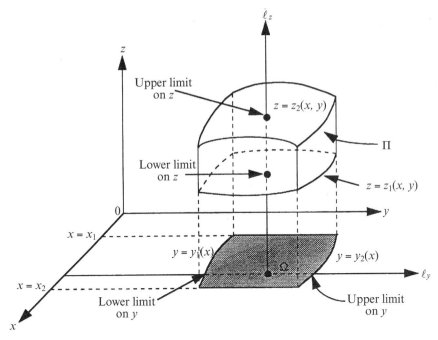

Fig. 7.16

obtain the limits we take a typical point (x, y) in the region Ω which is the two-dimensional 'shadow' of our three-dimensional region Π, and pass a vertical line ℓ_z through Π, from (x, y). The point where this line enters Π, $z = z_1(x, y)$, gives the lower z-limit, while the point where it leaves, $z = z_2(x, y)$, gives the upper limit. We now obtain the y-limits of integration by taking a typical value of x within the shadow and drawing a line ℓ_y parallel to the y-axis through x. The point $y = y_1(x)$ where this line enters the shadow gives the lower y-limit, the point where it leaves $y = y_2(x)$ gives the upper y-limit. Finally, we get the x-limits by choosing the extreme values of x which encompass the shadow.

Clearly, the triple integral over a region Π provides a simple expression for the volume of Π in the form

$$\text{Volume of } \Pi = \iiint_{\Pi} d\tau$$

PROBLEM 14

Find the volume of the solid enclosed within the paraboloids $z = x^2 + y^2$ and $z = 8 - x^2 - 3y^2$.

The solid Π and its shadow on the xy-plane, Ω, are illustrated in Fig. 7.17. To determine the shadow region Ω we find the curve of intersection of the paraboloids

$$z = x^2 + y^2 = 8 - x^2 - 3y^2$$

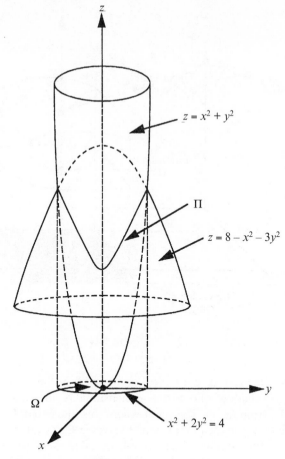

$z = x^2 + y^2$

Π

$z = 8 - x^2 - 3y^2$

$x^2 + 2y^2 = 4$

Ω

Fig. 7.17

or

$$x^2 + 2y^2 = 4$$

The curve of intersection is thus on the elliptic cylinder with this equation, which also defines the shadow region Ω.

Let (x, y) be a typical point in Ω. Then a vertical line through (x, y) enters the region Π at the surface $z = x^2 + y^2$ and leaves it at $z = 8 - x^2 - 3y^2$, which therefore define the lower and upper z-limits, respectively. At the same time a line through (x, y), parallel to the y-axis, passing through the shadow Ω gives lower y-limit $y = -(1/\sqrt{2})\sqrt{4-x^2}$ and upper y-limit $y = (1/\sqrt{2})\sqrt{4-x^2}$. Finally, x ranges between -2 and 2. Thus, the volume integral becomes

$$\text{Volume } \Pi = \int_{-2}^{2} \int_{-\frac{1}{\sqrt{2}}\sqrt{4-x^2}}^{\frac{1}{\sqrt{2}}\sqrt{4-x^2}} \int_{x^2+y^2}^{8-x^2-3y^2} dz\, dy\, dx$$

$$= \int_{-2}^{2} \int_{-\frac{1}{\sqrt{2}}\sqrt{4-x^2}}^{\frac{1}{\sqrt{2}}\sqrt{4-x^2}} (8 - 2x^2 - 4y^2) \, dy \, dx$$

$$= \int_{-2}^{2} \left[2(4 - x^2)y \frac{4}{3} - y^3 \right]_{-\frac{1}{\sqrt{2}}\sqrt{4-x^2}}^{\frac{1}{\sqrt{2}}\sqrt{4-x^2}} dx$$

$$= \frac{8\sqrt{2}}{3} \int_{0}^{2} (4 - x^2)^{3/2} \, dx$$

$$= 8\sqrt{2}\,\pi$$

Note that if we wish to change the order of integration in such integrals then we need to consider projections or 'shadows' on one of the other coordinate planes. In all there are six different possible orders of integration in this case.

Just as we considered double integrals in polar coordinates, so we can express triple integrals in different three-dimensional coordinate systems – the main problem being to find an expression for the volume element $d\tau$, and the integration limits.

● Cylindrical coordinates ─────────────────────

In this case the volume element is

$$d\tau = r \, dz \, dr \, d\theta$$

since it is simply the same as that for two-dimensional polar coordinates with the added 'vertical' z-dimension.

The triple integral in this case may therefore be written as

$$\iiint_{\tau} f \, d\tau = \int_{\theta_1}^{\theta_2} \int_{r_1(\theta)}^{r_2(\theta)} \int_{z_1(r,\,\theta)}^{z_2(r,\,\theta)} f(r, \theta, z) \, r \, dz \, dr \, d\theta$$

● Spherical coordinates ─────────────────────

The volume element in this case is

$$d\tau = \rho^2 (\sin \phi) \, d\rho \, d\phi \, d\theta$$

This may be obtained by direct geometric arguments, or using the methods of the next section – *see* Exercise 7.9.2.

The triple integral is therefore in this case:

$$\iiint_{\tau} f \, d\tau = \int_{\theta_1}^{\theta_2} \int_{\phi_1(\theta)}^{\phi_2(\theta)} \int_{\rho_1(\theta,\,\phi)}^{\rho_2(\theta,\,\phi)} f(\rho, \theta, \phi) \, \rho^2 (\sin \phi) \, d\rho \, d\phi \, d\theta$$

EXERCISES ON 7.8

1. Evaluate the triple integral

$$\iiint_P 4x^2yz^3 \, d\tau$$

where P is the rectangular box defined by

$$0 \le x \le 2, \qquad 0 \le y \le 3, \qquad -1 \le z \le 2$$

2. Evaluate $\iiint_\Pi z \, d\tau$, where Π is the wedge in the first octant cut from the cylindrical solid $y^2 + z^2 \le 1$ by the planes $y = x$ and $x = 0$.

3. Evaluate

$$\int_{-3}^{3} \int_{-\sqrt{9-x^2}}^{\sqrt{9-x^2}} \int_{0}^{9-x^2-y^2} (x^2 + y^2) \, dz \, dy \, dx$$

by conversion to cylindrical polars.

4. Find the mass of a sphere $x^2 + y^2 + z^2 = a^2$ if its density at any point is proportional to $k\rho$, where ρ is the distance of the point from the origin. (*Hint:* the mass is $\iiint_\Pi D \, d\tau$ where D is the density.)

7.9 Change of variables in multiple integrals: project

You may have noticed, particularly in the previous section, that when we 'changed variables' – from rectangular to polar coordinates, for example – we did not do so in the way familiar from ordinary integration. Rather than 'substitute' new variables, we essentially identify the region over which we are integrating, and re-represent it in terms of new variables. Thus, if the rectangular coordinate limits tell us we are dealing with a cylinder, then we re-express that cylinder in terms of cylindrical coordinates. This is all very well if we have simple geometric forms to deal with, but how do we handle general transformations between variables – analogous to substitution in ordinary integration? This section leads you to answer this yourself.

First, think what happens in ordinary integration when we make a substitution $x = x(u)$ to a new variable u in a definite integral in x. Provided $x(u)$ is differentiable, we obtain

$$\int_{a}^{b} f(x) \, dx = \int_{\alpha}^{\beta} f(x(u)) \left| \frac{dx(u)}{du} \right| du$$

where $\alpha < \beta$ are the limits on u appropriate to the x-interval $[a, b]$, and the modulus of the derivative is taken to allow for $x(u)$ being an increasing ($x'(u) > 0$) or decreasing ($x'(u) < 0$) function. Three things happen here – $f(x)$ is changed to a function of u, $f(x(u))$; the limits change by solving $a = x(u)$ and $b = x(u)$; and dx is replaced by $|x'(u)| \, du$. If we extend this idea to multiple integrals then the first two of these are fairly straightforward to generalize, but the third needs some work.

Specifically, consider the case of double integration, and suppose we wished to change x and y variables to, say, u and v by a transformation

$$x = x(u, v) \quad y = y(u, v)$$

What becomes of $dx\,dy$ in the transformation? If we think about polar coordinates, where $dx\,dy$ goes to $r\,dr\,d\theta$, then we would expect, in general, that $dx\,dy$ goes to something like $J(u, v)\,du\,dv$ where J (the notation is deliberate) is some function of u and v. But how do we find J? To answer this we have to think quite carefully about what the transformation does.

The *transformation* $x = x(u, v)$, $y = y(u, v)$ takes a typical point (u, v) of the uv-plane into the *image* (x, y) in the xy-plane, as shown in Fig. 7.18. Notice that we are representing u, v as rectangular coordinates, just like x, y – this is only a matter of pictorial convenience, it does not, for example, preclude the use of polar coordinates. Lines and curves in the uv-plane become mapped to different lines and curves in the xy-plane, and by a natural extension, any region in the uv-plane becomes mapped to some differently shaped region in the xy-plane. We are particularly interested in what happens to the sorts of elemental areas used in building up double integrals, so we look in particular at the transformation of an elementary uv-rectangle with sides given by lines $u = u_0$, $u = u_0 + \Delta u$, $v = v_0$, $v = v_0 + \Delta v_0$.

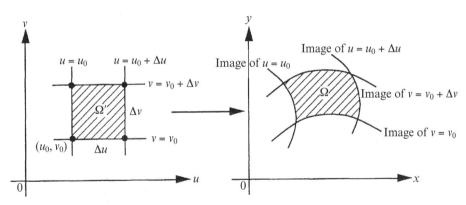

Fig. 7.18

Determine the region in the xy-plane which corresponds to the rectangle contained within $u = 1$, $u = 2$, $v = 1$, $v = 2$ in the uv-plane under the transformation

$$x = \sqrt{v/u} \quad y = \sqrt{uv}$$

Since $u = y/x$ and $v = xy$, the uv-lines go to

$$u = 1 \rightarrow y = x \qquad\qquad u = 2 \rightarrow y = 2x$$

$$v = 1 \rightarrow y = 1/x \qquad\qquad v = 2 \rightarrow y = 2/x$$

and the given rectangle is deformed to the curvilinear element contained within these rays and curves.

Now let

$$\mathbf{r} = \mathbf{r}(u, v) = x(u, v)\mathbf{i} + y(u, v)\mathbf{j}$$

be the position vector of the point (x, y) in the xy-plane corresponding to the point (u, v) in the uv-plane.

PROBLEM 16

Show that if Δu, Δv are small, the elementary uv-rectangle maps to a region in xy-space which can be approximated by a parallelogram with edges represented by vectors

$$\mathbf{a} = \mathbf{r}(u_0 + \Delta u, v_0) - \mathbf{r}(u_0, v_0)$$

$$\simeq \left(\frac{\partial x}{\partial u}\mathbf{i} + \frac{\partial y}{\partial u}\mathbf{j}\right)\Delta u$$

and

$$\mathbf{b} = \mathbf{r}(u_0, v_0 + \Delta v) - \mathbf{r}(u_0, v_0)$$

$$\simeq \left(\frac{\partial x}{\partial v}\mathbf{i} + \frac{\partial y}{\partial v}\mathbf{j}\right)\Delta v$$

with all partial derivatives evaluated at (u_0, v_0).

Deduce that the area of the xy-parallelogram thus defined is

$$\Delta A \simeq \left(\frac{\partial x}{\partial u}\frac{\partial y}{\partial v} - \frac{\partial x}{\partial v}\frac{\partial y}{\partial u}\right)\Delta u\,\Delta v$$

$$= \begin{vmatrix} \dfrac{\partial x}{\partial u} & \dfrac{\partial x}{\partial v} \\[2mm] \dfrac{\partial y}{\partial u} & \dfrac{\partial y}{\partial v} \end{vmatrix} \Delta u\,\Delta v$$

The use of the determinantal notation here is deliberate – it allows a natural extension to triple and multiple integrals.

The object

$$J(u, v) \equiv \begin{vmatrix} \dfrac{\partial x}{\partial u} & \dfrac{\partial x}{\partial v} \\[2mm] \dfrac{\partial y}{\partial u} & \dfrac{\partial y}{\partial v} \end{vmatrix}$$

is precisely the scaling factor referred to earlier. It is called the *Jacobian* for the transformation $x = x(u, v)$, $y = y(u, v)$, named after the German mathematician Carl Gustav Jacob Jacobi (1804–51).

The Jacobian is sometimes written in the form

$$J(u, v) = \left| \frac{\partial(x, y)}{\partial(u, v)} \right|$$

for reasons soon to become obvious.

PROBLEM 17

If the transformation $x = x(u, v)$, $y = y(u, v)$ maps the region Ω' in the uv-plane into the region Ω in the xy-plane, and the Jacobian does not change sign or vanish on Ω', then show that, subject to appropriate good behaviour,

$$\iint_{\Omega} f(x, y) \, dx \, dy = \iint_{\Omega'} f(x(u, v), y(u, v)) \left| \frac{\partial(x, y)}{\partial(u, v)} \right| \, du \, dv$$

Note the similarity in form to the single-variable case.

The proof amounts to obtaining the integrals on both sides by totting up the respective contributions from the elemental areas which make up the regions Ω and Ω', using the result of Problem 16.

The result of Problem 17 now enables us to change variables in a double integral by a routine procedure. Such substitutions may be chosen to ease the integration, by simplifying either the integrand or the limits.

PROBLEM 18

Evaluate

$$\iint_{\Omega} xy \, dx \, dy$$

where Ω is the region in the first quadrant bounded by
$$x^2 - y^2 = 1, \qquad x^2 - y^2 = 4, \qquad x^2 + y^2 = 9, \qquad x^2 + y^2 = 16$$

The boundaries of this region are curved; it seems natural to make a transformation which straightens them out, so we take

$$u = x^2 + y^2 \qquad v = x^2 - y^2$$

The Jacobian is

$$J(u, v) = \frac{-1}{4\sqrt{u^2 - v^2}}$$

and the transformed integral becomes

$$\iint_{\Omega} xy \, dx \, dy = \iint_{\Omega'} \frac{1}{8} \, du \, dv = \frac{21}{8}$$

PROBLEM 19

Extend the above results to triple integrals.

This is a good illustration of the power of the right notation when extending results from lower to higher dimensions.

If the transformation $x = x(u, v, w)$, $y = y(u, v, w)$, $z = z(u, v, w)$ maps the region Π' in uvw-space to the region Π in xyz-space, and if the Jacobian

$$\frac{\partial(x, y, z)}{\partial(u, v, w)} = \begin{vmatrix} \dfrac{\partial x}{\partial u} & \dfrac{\partial x}{\partial v} & \dfrac{\partial x}{\partial w} \\[2mm] \dfrac{\partial y}{\partial u} & \dfrac{\partial y}{\partial v} & \dfrac{\partial y}{\partial w} \\[2mm] \dfrac{\partial z}{\partial u} & \dfrac{\partial z}{\partial v} & \dfrac{\partial z}{\partial w} \end{vmatrix}$$

does not change sign or vanish on Π, then modulo appropriate good behaviour,

$$\iiint_\Pi f(x, y, z)\, dx\, dy\, dz = \iiint_{\Pi'} f(x(u, v, w), y(u, v, w), z(u, v, w)) \left| \frac{\partial(x, y, z)}{\partial(u, v, w)} \right| du\, dv\, dw$$

Now, an interesting question. Does this not seem too good to be true? You can almost guess what is going to happen in higher dimensions – what is the underlying mathematical theory that leads to such 'simple' results? We cannot imagine generalizing Problem 16 to higher dimensions, so what do we do? In fact, we require a complete generalization of vector calculus to *tensor calculus*, and the theory of *differential forms*. These provide tools which lead very naturally to the results of this section, but are beyond the scope of this book.

FURTHER EXERCISES

1. Integrate the following functions over the rectangles with vertices (a) (0, 0), (2, 0), (2, 1), (0, 1), (b) (–1, –1), (0, –1), (0, 2), (–1, 2):

   (i) xy (ii) $2x + y$ (iii) $3x^2 y$ (iv) $(x + y)e^{xy}$ (v) $x^2 y + 3y^3$

2. Find the volume above the rectangle $\{(x, y) : -1 \le x \le 0, 1 \le y \le 2\}$ and under the surface

   (i) $z = x + y$ (ii) $z = x^3 y + x$ (iii) $z = x^2 + y^2$ (iv) $z = x^2 e^y$

3. Let Ω be the disc $\{(x, y) : x^2 + y^2 \le 1\}$. Find lower and upper bounds for

   $$\iint_\Omega \frac{1}{1 + x^2 + y^2}\, dx\, dy$$

4. Evaluate the double integrals of the following functions over the regions given:

   (i) $z = x^2 + y^2$ $\{(x, y) : 0 \le x \le 1, x^2 \le y \le \sqrt{x}\}$
   (ii) $z = 3x + 2y$ $\{(x, y) : 0 \le x \le 2, x^2 \le y \le 2x\}$
   (iii) $z = x^4 - 2y$ $\{(x, y) : -1 \le x \le 1, -x^2 \le y \le x^2\}$

(iv) $z = xy$ $\quad\quad\quad \{(x, y) : 2 \le x \le 4, \frac{1}{2}x \le y \le \sqrt{x}\}$

(v) $z = 2x - y^2$ $\quad\quad \{(x, y) : 1 - y \le x \le y - 1, 1 \le y \le 3\}$
(vi) $z = 1/x$ $\quad\quad\quad \{(x, y) : y^6 \le x \le 3y^6, 1 \le y \le 2\}$
(vii) $z = 2 - x - y$ $\quad \{(x, y) : y \le x \le 1, 0 \le y \le 1\}$
(viii) $z = xy + y^2$ $\quad\quad \{(x, y) : -1 \le x \le y, 0 \le y \le 1\}$

5. Evaluate the following integrals in the order given and in reverse order:

(i) $\displaystyle\int_0^1 \int_0^x (3 - x - y)\, dy\, dx$

(ii) $\displaystyle\int_0^1 \int_{x^2}^{x^{1/4}} (x^{1/2} - y^2)\, dy\, dx$

(iii) $\displaystyle\int_0^1 \int_1^{e^x} dy\, dx$

(iv) $\displaystyle\int_1^2 \int_1^{x^2} \frac{x}{y}\, dy\, dx$

6. For the following sketch the region of integration and evaluate the integral by choosing the best order of integration:

(i) $\displaystyle\int_0^{\pi} \int_x^{\pi} \frac{\sin y}{y}\, dy\, dx$

(ii) $\displaystyle\int_0^{2\sqrt{\ln 2}} \int_{y/2}^{\sqrt{\ln 2}} e^{x^2}\, dx\, dy$

(iii) $\displaystyle\int_0^2 \int_0^{4 - x^2} \frac{xe^{2y}}{4 - y}\, dy\, dx$

(iv) $\displaystyle\int_0^8 \int_{x^{1/3}}^2 \frac{dx\, dy}{y^4 + 1}$

7. Find the volume under the surface and above the region given:
   (i) $z = xy^2 + 2y^3$ $\quad \{(x, y) : 0 \le x \le 2, 1 \le y \le 3\}$
   (ii) $z = 2x - y^2$ $\quad$ the triangle bounded by $y = -x + 1, y = x + 1, y = 3$
   (iii) $z = x^2 + y^2$ $\quad$ the triangle bounded by $y = x, x = 0, x + y = 2$
   (iv) $z = x + 4$ $\quad$ the region bounded by the parabola $y = 4 - x^2$ and the line $y = 3x$.

8. Evaluate the improper integrals

(i) $\displaystyle\int_1^{\infty} \int_{e^x}^1 \frac{1}{x^3 y}\, dy\, dx$

(ii) $\displaystyle\int_{-\infty}^{\infty} \int_{-\infty}^{\infty} \frac{1}{(x^2 + 1)(y^2 + 1)}\, dx\, dy$

9. Evaluate the following integrals by converting to polar coordinates:

(i) $\displaystyle\int_{-1}^1 \int_0^{\sqrt{1-x^2}} dy\, dx$

(ii) $\displaystyle\int_{-a}^a \int_{-\sqrt{a^2-x^2}}^{\sqrt{a^2-x^2}} dy\, dx$

(iii) $\displaystyle\int_{-1}^0 \int_{-\sqrt{1-x^2}}^0 \frac{2}{1 + \sqrt{x^2 + y^2}}\, dy\, dx$

(iv) $\displaystyle\int_0^2 \int_0^{\sqrt{1-(x-1)^2}} \frac{x + y}{x^2 + y^2}\, dy\, dx$

10. Find the area enclosed by one leaf of the rose $r = 4 \cos 2\theta$.

11. Find the area of the region cut from the first quadrant by the cardioid $r = 1 + \sin \theta$.

12. Evaluate the integral $\int_0^\infty e^{-x^2} dx$ using double integration.

13. Evaluate the following triple integrals:

    (i) $\int_0^1 \int_0^1 \int_0^1 (x^2 + y^2 + z^2) \, dz \, dy \, dx$

    (ii) $\int_1^e \int_1^e \int_1^e \frac{1}{xyz} \, dx \, dy \, dz$

    (iii) $\int_0^\pi \int_0^\pi \int_0^\pi \cos(u + v + w) \, du \, dv \, dw$

    (iv) $\int_0^7 \int_0^2 \int_0^{\sqrt{4-q^2}} \frac{q}{r+1} \, dp \, dq \, dr$

14. Find the volumes of the following regions:
    (i) The region between the cylinder $z = y^2$ and the xy-plane that is bounded by the planes $x = 0$, $x = 1$, $y = -1$, $y = 1$.
    (ii) The tetrahedron with vertices at the points $(0, 0, 0)$, $(1, 0, 0)$, $(0, 1, 0)$, $(0, 0, 1)$.

15. Use double integration to calculate the area that lies between the curves
    (i) $\sqrt{x} + \sqrt{y} = \sqrt{a}$ and $x + y = a$, $a > 0$
    (ii) $y = x$ and $y = x^2$ in the first quadrant.

16. Sketch the region of integration and change the order of integration:

    (i) $\int_0^1 \int_{x^4}^{x^2} f(x, y) \, dy \, dx$

    (ii) $\int_0^1 \int_{-y}^{y} f(x, y) \, dx \, dy$

17. Evaluate

    (i) $\int_{0}^{2} \int_{-1}^{1} \int_{1}^{3} (z - xy) \, dy \, dx \, dz$

    (ii) $\int_0^{\pi/2} \int_0^1 \int_0^{\sqrt{1-x^2}} x(\cos z) \, dy \, dx \, dz$

18. Evaluate

    (i) $\int_0^\pi \int_0^2 \int_0^{4-r^2} r \, dz \, dr \, d\theta$

    (ii) $\int_0^{\pi/4} \int_0^1 \int_0^{r\cos\theta} r^2(\sec^3\theta) \, dz \, dr \, d\theta$

19. Find the volume of the solid bounded above by the cone $z^2 = x^2 + y^2$, below by the xy-plane, and on the sides by the hemisphere $z = \sqrt{4 - x^2 - y^2}$.

20. Evaluate the repeated integrals

    (i) $\int_0^{\pi/3} \int_0^{2\pi} \int_0^1 \rho^2 \sin\phi \, d\rho \, d\theta \, d\phi$

    (ii) $\int_0^{\pi/4} \int_0^\pi \int_0^{2\cos\phi} \rho^2 \sin\phi \, d\rho \, d\theta \, d\phi$

21. Derive the formula for the volume of a sphere using spherical coordinates.

22. Evaluate the volume integral of the function $f(x, y, z) = z$ over the region Π bounded by the surfaces

$$z = \sqrt{x^2 + y^2}, \ x = 0, \ x - y = 0 \text{ and } z = 4$$

23. Find the Jacobians for the following transformations:
    (i) $x = au + bv$, $y = cu + dv$
    (ii) $x = \sin u + \cos v$, $y = -\cos u + \sin v$
    (iii) $x = uvw$, $y = u - uw$, $z = vw$
    (iv) $x = \cos s \cos t$, $y = \cos r \cos s$, $z = \cos r \cos t$

24. For the following transformations verify that

$$\frac{\partial(x, y)}{\partial(u, v)}\frac{\partial(u, v)}{\partial(x, y)} = 1$$

    (i) $x = uv$, $y = v - uv$
    (ii) $x = v^2/u$, $y = v/u$

25. Use the result for the product of two 2×2 determinants and the chain rule to prove the result stated in Exercise 24.

26. Use the transformation $u = xy$, $v = xy^4$ to evaluate

$$\iint_\Omega \sin(xy) \, dx \, dy$$

where Ω is the region enclosed by the curves $y = \pi/x$, $y = 2\pi/x$, $xy^4 = 1$, $xy^4 = 2$.

27. Evaluate

$$\iint_\Omega \sin(x - y) \cos(x + 2y) \, dx \, dy$$

where Ω is the parallelogram enclosed by $x - y = 0$, $x - y = \pi$, $x + 2y = 0$, $x + 2y = \pi/2$, by using an appropriate transformation.

28. Essay and discussion topics:
    (i) Difficulties encountered in going from single to multiple integration.
    (ii) The integral as an area extended to double and triple integrals.
    (iii) Fubini's theorem and change of order of integration.
    (iv) Finding limits of integration in multiple integration.
    (v) Transformation of variables in multiple integration.
    (vi) Applications of multiple integration.
    (vii) Lessons from the life of Jacobi.

$8 \bullet$ Functions of a Vector

8.1 Introduction: what is a vector?

You are probably aware from your previous mathematics that one of the most powerful tools for dealing with functions of more than one variable is the use of vector notation. Thus, when describing the motion of a projectile in two dimensions we have the option of using $(x(t), y(t))$ coordinates to represent position at any time or of using the *position vector*

$$\mathbf{r}(t) = x(t)\mathbf{i} + y(t)\mathbf{j}$$

with $\mathbf{i}, \mathbf{j}$ the usual basis vectors. The point about using vectors here is that it effectively reduces the number of 'variables' – instead of two coordinates, we have one vector.

A natural question therefore is to what extent one can use vectors in the theory of functions of several variables – which is what this book is about. So far, we have managed virtually without them – and indeed it is not always clear that they would be of much use. But in fact they are – the bulk of the rest of this book is essentially about 'vector calculus'. Why should this be so? Why does most of the differentiation and integration that we cover occur in a vector context? Is it just the case that vectors are simply a good shorthand notation whenever we have more than one variable? No, there is a deep significance to the idea of a vector which is simply not brought out in the introductory treatments to which you may have been exposed so far. Vectors are of fundamental importance and utility in all physical applications – they are far more than simply a shorthand notation. The reason for this is rather subtle, and requires a leap into abstraction which really has to be deferred until sufficient mathematical foundations have been laid – that is why elementary treatments are usually incomplete. To get an insight into the new viewpoint, let us stick with our two-dimensional projectile.

PROBLEM I

Define what is meant by a vector.

You may have a number of different definitions, but they are most likely to come from the following list:

1. Any quantity having both a magnitude and a direction, such as velocity as opposed to speed.
2. An n-tuple of real numbers viewed as a member of an n-dimensional Euclidean space.
3. An element of a vector space.

Definition 1 is the one we meet at the most elementary level. Certainly, our position vector $\mathbf{r}(t)$ seems to fit this description, if we think of it in the traditional way as a 'directed line segment' – an 'arrow'. But there is more to it than that. The representation as a vector $\mathbf{r}(t)$ does more than replace x, y by a single vector – it 'hides' the coordinate system altogether. If we choose a different coordinate system – say, by rotating the x- and y-axes, as we might do if considering motion on an inclined plane – then this would only change the *representation* of $\mathbf{r}(t)$ in terms of the coordinate system; it would not change $\mathbf{r}(t)$ itself.

PROBLEM 2

> Consider a rotation about the origin of the x- and y-axes through an angle θ in an anti-clockwise sense to a new set of axes x', y'. Express the coordinates (x', y') of a point P referred to the x'- and y'-axes in terms of the coordinates (x, y) referred to the x- and y-axes.

If we regard (x, y) as the components of a vector $\mathbf{r} = \overrightarrow{OP}$ relative to the x- and y-axes and (x', y') as the components of the same vector relative to the x'-, y'-axes, then it is a standard exercise in coordinate geometry to show that under the rotation of axes the components of $\mathbf{r}$ transform according to

$$x' = (\cos \theta)x + (\sin \theta)y$$
$$y' = -(\sin \theta)x + (\cos \theta)y$$

Any quantity whose components transform in this way is called a *two-dimensional vector*. This is the key to the general definition of a vector – its behaviour under a particular transformation of the coordinate system. So the vectorial quality is really a reflection of the way a coordinate system to which the vector is referred transforms. In the particular case considered here – the rotation of rectangular Cartesian coordinates – we say that the vector is a *Cartesian vector*.

A *scalar* quantity, α, represented in terms of such transformations of coordinate systems is one which transforms according to

$$\alpha' = \alpha$$

under a rotation of axes.

These simple two-dimensional examples show that we need to look at transformations of coordinate systems – in particular, *rotations* of rectangular coordinate systems.

8.2 Rotation of axes

Consider a three-dimensional rectangular coordinate system with origin 0 and axes $0x$, $0y$, $0z$. Suppose that it is rotated about 0 to give another set of axes $0x'$, $0y'$, $0z'$. Let the direction cosines of the axis $0x'$ relative to the axes $0x$, $0y$, $0z$ be ℓ_{11}, ℓ_{12}, ℓ_{22}, respectively. Similarly, the direction cosines of $0y'$, $0z'$ are denoted by ℓ_{21}, ℓ_{22}, ℓ_{23} and ℓ_{31}, ℓ_{32}, ℓ_{33}, respectively. We may represent this in an array

0	x	y	z
x'	ℓ_{11}	ℓ_{12}	ℓ_{13}
y'	ℓ_{21}	ℓ_{22}	ℓ_{23}
z'	ℓ_{31}	ℓ_{32}	ℓ_{33}

Reading vertically, we get the direction cosines of the $0x$, $0y$, $0z$ axes with respect to the $0x'$, $0y'$, $0z'$. Reading horizontally gives the direction cosines of the $0x'$, $0y'$, $0z'$ axes with respect to the $0x$, $0y$, $0z$ axes.

Since the $0x'$, $0y'$, $0z'$ axes are mutually perpendicular we have the following *orthogonality conditions* (why?):

$$\begin{aligned}
\ell_{11}\ell_{21} + \ell_{12}\ell_{22} + \ell_{13}\ell_{23} &= 0 \\
\ell_{21}\ell_{31} + \ell_{22}\ell_{32} + \ell_{23}\ell_{33} &= 0 \\
\ell_{31}\ell_{11} + \ell_{32}\ell_{12} + \ell_{33}\ell_{13} &= 0
\end{aligned} \tag{8.1}$$

Also, from the properties of direction cosines we have

$$\begin{aligned}
\ell_{11}^2 + \ell_{12}^2 + \ell_{13}^2 &= 1 \\
\ell_{21}^2 + \ell_{22}^2 + \ell_{23}^2 &= 1 \\
\ell_{31}^2 + \ell_{32}^2 + \ell_{33}^2 &= 1
\end{aligned} \tag{8.2}$$

The relations (8.1) and (8.2) together are called the *orthonormality relations* for the direction cosines.

Using the symmetry between the dashed and undashed sets of axes – the elements of the columns of the ℓ-array are the direction cosines of the axes $0x$, $0y$, $0z$ relative to the axes $0x'$, $0y'$, $0z'$ – we obtain an alternative form for the orthonormality relations:

$$\begin{aligned}
\ell_{11}\ell_{12} + \ell_{21}\ell_{22} + \ell_{31}\ell_{32} &= 0 \\
\ell_{12}\ell_{13} + \ell_{22}\ell_{23} + \ell_{32}\ell_{33} &= 0 \\
\ell_{13}\ell_{11} + \ell_{23}\ell_{21} + \ell_{33}\ell_{31} &= 0
\end{aligned} \tag{8.3}$$

$$\begin{aligned}
\ell_{11}^2 + \ell_{21}^2 + \ell_{31}^2 &= 1 \\
\ell_{12}^2 + \ell_{22}^2 + \ell_{32}^2 &= 1 \\
\ell_{13}^2 + \ell_{23}^2 + \ell_{33}^2 &= 1
\end{aligned} \tag{8.4}$$

We now define a *transformation matrix*

$$L = \begin{bmatrix} \ell_{11} & \ell_{12} & \ell_{13} \\ \ell_{21} & \ell_{22} & \ell_{23} \\ \ell_{31} & \ell_{32} & \ell_{33} \end{bmatrix}$$

from the array of direction cosines. Using the orthonormality relations it is easy to see that, with L^{T} denoting the transpose of L,

$$LL^{\mathrm{T}} = L^{\mathrm{T}}L = I$$

where I is the unit matrix. Taking the determinant gives

$$|LL^{\mathrm{T}}| = |L||L^{\mathrm{T}}| = |L^2| = 1$$

so

$$|L| = \pm 1$$

This is a characteristic property of any *orthogonal transformation* (*see*, for example,

Anton and Rorres, 1991). Now when the dashed and undashed axes coincide ($0x = 0x'$, etc.) we have $\ell_{ii} = 1$ and $\ell_{ij} = 0$, $i \neq j$, for all i and j and so L then becomes the unit matrix, with $|L| = 1$. So for strict rotations we always have $|L| = 1$ ($|L| = -1$ corresponds to a reversal of axes).

So, to summarize, a necessary and sufficient condition that L represents a rotation of axes is that it be orthogonal with unit determinant, i.e.

$$LL^T = I = L^TL$$

$$|L| = 1$$

With these conditions the rows of the transformation matrix represent direction cosines of $0x'$, $0y'$, $0z'$ with respect to $0x$, $0y$, $0z$, and similarly for the columns.

The above describes the relationship between rotated axes. An obvious question concerns the relationship between the coordinates (x', y', z') of a point P referred to the dashed axes and the corresponding coordinates (x, y, z) of the point P with respect to the undashed axes. By the definition of the direction cosines applied to the orthogonal projections of $0P$ on the $0x'$, $0y'$, $0z'$ axes, we obtain

$$x' = \ell_{11} x + \ell_{12} y + \ell_{13} z$$
$$y' = \ell_{21} x + \ell_{22} y + \ell_{23} z$$
$$z' = \ell_{31} x + \ell_{32} y + \ell_{33} z$$

which shows how the coordinates of P transform under a rotation of axes. If we write the coordinates as column vectors $\mathbf{x}$, $\mathbf{x}'$ then this transformation can be written in matrix form

$$\mathbf{x}' = L\mathbf{x}$$

Since $L^TL = I$, we can easily derive the inverse transformation:

$$\mathbf{x} = L^T\mathbf{x}'$$

or

$$x = \ell_{11} x' + \ell_{21} y' + \ell_{31} z'$$
$$y = \ell_{12} x' + \ell_{22} y' + \ell_{32} z'$$
$$z = \ell_{13} x' + \ell_{23} y' + \ell_{33} z'$$

EXERCISES ON 8.2

1. The axes $0x'y'z'$, initially coincident with the axes $0xyz$, are rotated through an angle θ about the z-axis, in the direction from the x-axis to the y-axis. Show that

$$x' = x \cos \theta + y \sin \theta$$
$$y' = -x \sin \theta + y \cos \theta$$
$$z' = z$$

2. Show that the transformation defined by

$$x' = \frac{1}{\sqrt{2}} x - \frac{1}{\sqrt{6}} y + \frac{1}{\sqrt{3}} z$$

$$y' = \frac{1}{\sqrt{2}} x + \frac{1}{\sqrt{6}} y - \frac{1}{\sqrt{3}} z$$

$$z' = \frac{2}{\sqrt{6}} y + \frac{1}{\sqrt{3}} z$$

corresponds to a rotation of axes.

8.3 The summation convention

If you have not already tired of writing out all these ℓ_{ij} expressions, then you soon will. Relief is at hand, however, in the form of a useful convention for representing summations of indexed quantities. This is almost essential in the study of an important extension of vectors, called *tensors*, and their treatment in n-dimensional spaces. Once grasped, the convention is incredibly powerful to use, and reduces long tedious calculations to elementary operations. First we change the notation for the coordinates to subscript notation, replacing (x, y, z) by (x_1, x_2, x_3) and similarly (x', y', z') by (x'_1, x'_2, x'_3).

PROBLEM 3

Rewrite the transformation equations using this notation and note any patterns you observe in the result.

We obtain

$$x'_1 = \ell_{11} x_1 + \ell_{12} x_2 + \ell_{13} x_3$$
$$x'_2 = \ell_{21} x_1 + \ell_{22} x_2 + \ell_{23} x_3$$
$$x'_3 = \ell_{31} x_1 + \ell_{32} x_2 + \ell_{33} x_3$$

The most obvious thing here is the occurrence of repeated subscripts, as in $\ell_{12} x_2$, etc. Indeed, we can write the equations as

$$x'_i = \sum_{j=1}^{3} \ell_{ij} x_j \qquad i = 1, 2, 3$$

Similarly, the inverse transformation can be written

$$x_i = \sum_{j=1}^{3} \ell_{ji} x'_j \qquad i = 1, 2, 3$$

This is already a considerable simplification, of course, but perhaps you will notice one further step – we do not need the summation signs at all, or the restriction on the number of values of the subscripts (three in this case) if we adopt the convention that *repeated subscripts are always summed over their range of values*. This is the *summation convention*. It enables us to write the above equations as

$$x'_i = \ell_{ij} x_j$$
$$x_i = \ell_{ji} x'_j$$

The 'free' unrepeated index, i, is assumed to range over all its values, while the repeated index, j, is summed over all its values.

The utility of this notation is greatly increased by introducing the *Kronecker delta* (Cox, 1996)

$$\delta_{ij} = \begin{cases} 0 & \text{when } i \neq j \\ 1 & \text{when } i = j \end{cases}$$

Using this and the summation convention, the orthonormality relations can now be written

$$\ell_{ik}\,\ell_{jk} = \delta_{ij}$$

or in the alternative form

$$\ell_{ki}\,\ell_{kj} = \delta_{ij}$$

PROBLEM 4

Show that $\delta_{ij}\,x_j = x_i$.

We have

$$\delta_{ij}\,x_j = \delta_{i1}\,x_1 + \delta_{i2}\,x_2 + \delta_{i3}\,x_3$$

and since all δ_{ij} are zero except δ_{ii}, which equals 1, the only term remaining on the right-hand side is x_i, as required.

PROBLEM 5

Use the new notation and conventions to deduce the inverse transformation to $x_i' = \ell_{ij}\,x_j$.

We want to express the x_i in terms of the x_i'. 'Multiply' both sides of the transformation by ℓ_{ki} to give

$$\ell_{ki}\,x_i' = \ell_{ki}\,\ell_{ij}\,x_j$$

While this looks like a multiplication by ℓ_{ki}, it is in fact a linear combination, sometimes called a *contraction* – for example,

$$\ell_{ki}\,x_i' = \ell_{k1}\,x_1' + \ell_{k2}\,x_2' + \ell_{k3}\,x_3'$$

Using the orthonormality relations we have $\ell_{ki}\,\ell_{ij} = \ell_{ik}\,\ell_{ij} = \delta_{kj}$, so

$$\ell_{ki}\,x_i' = \delta_{kj}\,x_j = x_k$$

So the inverse transformation is

$$x_i = \ell_{ij}\,x_j'$$

where we have changed the 'dummy indices'.

1. Let a_{ij} be a quantity defined by

$$
\begin{array}{lll}
a_{11} = 1 & a_{12} = 0 & a_{13} = -1 \\
a_{21} = 3 & a_{22} = 1 & a_{23} = 2 \\
a_{31} = 2 & a_{32} = -1 & a_{33} = 0
\end{array}
$$

and let b_i be the components of a vector given by

$$b_1 = 4 \qquad b_2 = -1 \qquad b_3 = 1$$

Evaluate:

(i) $a_{1j}a_{j2}$ (ii) $a_{k1}a_{2k}$ (iii) $\delta_{1j}a_{j3}$

(iv) $a_{3j}a_{j1}$ (v) $a_{1j}b_j$ (vi) $\delta_{2j}a_{ji}b_i$

(vii) $a_{1j}a_{3k}b_jb_k$

2. Express the following using the summation convention

(i) $d\phi = \dfrac{\partial\phi}{\partial x_1}dx_1 + \dfrac{\partial\phi}{\partial x_2}dx_2 + \ldots + \dfrac{\partial\phi}{\partial x_N}dx_N$

(ii) $x_1^2 + x_2^2 + \ldots + x_N^2$

(iii) $a_{j1}b_1 + a_{j2}b_2 + \ldots + a_{jN}b_N$

8.4 Definition of a vector

You know that vectors are used to represent physical quantities such as velocity, force, and so on. There are of course plenty of other physical quantities, such as temperature and kinetic energy, which are not vectors – they are represented by ordinary numbers, or what we loosely call 'scalars' . We glibly say 'scalars are pure numbers having only a magnitude, while vectors are quantities having both a magnitude *and a direction*'. But is this all there is to it? Strain in a piece of metal also has a magnitude and a 'directional quality' – but can it be represented by a vector? No, as it happens. So, what is the essence of a vector?

The clue lies in the fact that scalars and vectors are used to represent physical quantities of a certain type, and are used to express various physical laws in a succinct form (such as Newton's laws of motion). In the practical implementation of these laws we have to choose a coordinate system to which we refer the scalar and vector quantities. We can choose an infinite variety of such coordinate systems – even in the case of two-dimensional rectangular coordinates we can translate the origin or rotate the axes. But – and this is the crucial thing – the laws of physics, and its fundamental concepts, do not change if we change the coordinate system. Their mathematical *representation* might change, but the laws do not. This suggests that scalars and vectors are objects which behave in a certain convenient way under transformations of coordinate systems. Since there are many types of coordinate systems, we expect there to be many types of 'vectors'. In fact, the elementary vectors with which we are most familiar are vectors with respect to *translations and rotations of rectangular axes*, since these are the main operations one has to allow for in three-dimensional classical mechanics. In relativity theory, or in elementary particle physics, we have to allow for much wider classes of transformation (the Lorentz transformation in special relativity, for example). In such cases we can define 'vectors' in an analogous way to what we are about to do for translations and rotations (indeed, you may have seen such phrases as 'rotations of space-time' in the popular science literature). Another point arising from our strain example given above is that there are other objects, more complicated than vectors, that are also used to represent physical quantities. The same arguments apply to them – they must be objects which behave in a certain way under coordinate transformations. These are the 'tensors' we referred to earlier.

The precise definition of a vector which we are about to give is designed for a specific mathematical purpose, and is the sense in which the term 'vector' is used throughout this book. Elsewhere the abuse of the term is rife. 'Vectors' in sociology, psychology, etc., are not vectors in our sense – there is no coordinate system, and they are simply lists of numbers, with no particular transformational properties. So, at long last to the definition, which, hopefully, will now be less of a shock!

A (Cartesian) *vector* is any mathematical object which satisfies the following conditions:

(i) When referred to a set of rectangular Cartesian axes $0xyz$ it can be represented uniquely by three numbers a_1, a_2, a_3 on the axes $0x$, $0y$, $0z$ respectively.

(ii) The triple of numbers is invariant under a translation of the axes, i.e. if $0xyz$, $0'x'y'z'$ are rectangular Cartesian axes with different origins 0, $0'$, such that $0x$ is parallel to $0'x'$, $0y$ is parallel to $0'y'$ and $0z$ is parallel to $0'z'$ and if the triads associated with the two coordinate systems are (a_1, a_2, a_3) and (a'_1, a'_2, a'_3) respectively, then

$$a_1 = a'_1 \qquad a_2 = a'_2 \qquad a_3 = a'_3$$

(iii) If the triples of numbers associated with two sets of axes $0xyz$, $0x'y'z'$ with the same origin 0 are (a_1, a_2, a_3), (a'_1, a'_2, a'_3) respectively, and if the direction cosines of $0x'$, $0y'$, $0z'$ relative to the axes $0xyz$ are given by the transformation matrix L in Section 8.2, then

$$a'_1 = \ell_{11} a_1 + \ell_{12} a_2 + \ell_{13} a_3$$
$$a'_2 = \ell_{21} a_1 + \ell_{22} a_2 + \ell_{23} a_3$$
$$a'_3 = \ell_{31} a_1 + \ell_{32} a_2 + \ell_{33} a_3$$

or

$$a'_i = \ell_{ij} a_j \qquad i = 1, 2, 3 \tag{8.5}$$

This rather lengthy definition can essentially be reduced to defining a vector as a triple of numbers

$$\mathbf{a} = (a_1, a_2, a_3)$$

where the components a_i are coordinates with respect to a rectangular Cartesian coordinate system, which transform according to (8.5) under rotation of axes (the translation invariance is usually taken as read). Problem 2 provides a simple two-dimensional example which may help here. Henceforth, we will always assume, unless otherwise stated, that the triple (a_1, a_2, a_3) refers to axes $0xyz$. Later in the book we will use the above definitions to confirm that certain new objects that we meet – grad, curl, etc. – really are vectors. Apart from this, the precise definition does not really impinge very much on the use of vectors. The point is that vectors allow us not only to represent two or more dimensions in terms of a single vector object, but also to do it in a way that is convenient for the representation of physical laws. So far as the methodology of vectors is concerned, I will assume a background roughly along the lines of Hirst (1995).

What about scalars? They, too, are related to physical quantities and must therefore behave sensibly under coordinate transformations. In fact, (Cartesian) scalars are objects which are *invariant* under translations and rotations of the objects – they are represented by pure numbers.

PROBLEM 6

> Prove that the kinetic energy $\frac{1}{2}mv^2 = \frac{1}{2}m\mathbf{v} \cdot \mathbf{v}$ is a scalar under rotation of the axes.

Of course, that the kinetic energy is formed from the 'scalar' product of two vectors gives some reassurance as to its 'scalar' behaviour here, but we need a more rigorous verification than that. Note that m is an ordinary number, unaffected by any change in the coordinate system – only the velocity, which is a vector, will be affected. Under a rotation of axes, from our definition above, it will transform according to

$$v_i' = \ell_{ij} v_j$$

So, in the rotated coordinate system the kinetic energy is given by

$$\frac{1}{2}mv_i' v_i' = \frac{1}{2}m\ell_{ij} v_j \, \ell_{ik} v_k = \frac{1}{2}m(\ell_{ij} \ell_{ik})v_j v_k$$

$$= \frac{1}{2}m\delta_{jk}v_j v_k = \frac{1}{2}mv_j v_j$$

using orthonormality.

This example illustrates the great power of our new notation and the summation convention for vectors. It tells us that the expression for the kinetic energy is identical in form and value in any rectangular coordinate systems related by rotations.

In general, consider a function $f(\alpha_1, \alpha_2, \dots)$ of several quantities $\alpha_1, \alpha_2, \dots$, such that given any set of rectangular axes $0xyz$, the elements $\alpha_1, \alpha_2, \dots$ are determined by a definite rule. Denote by $\alpha_1', \alpha_2', \dots$ the quantities corresponding to any other set of rectangular axes $0x'y'z'$ with the same origin. Then if

$$f(\alpha_1', \alpha_2', \dots) = f(\alpha_1, \alpha_2, \dots)$$

we say the function f is *invariant with respect to a rotation of the axes* – i.e. is a scalar relative to the transformation.

We now want to build up a complete 'calculus' of vectors. We want to define functions of vectors and study how to differentiate and integrate them. This is the subject of the rest of the book. In this chapter we look at vector functions of a single variable (such as $\mathbf{r}(t)$ in Section 8.1) and how they can be used to represent curves in space.

1. Show that $x_i x_i \equiv x_1^2 + x_2^2 + x_3^2$ is invariant under a rotation of axes.

2. The transformation matrix for a rotation from axes $0xyz$ to axes $0x'y'z'$ is given by

$$L = \begin{bmatrix} 0 & 1 & 0 \\ -1 & 0 & 0 \\ 0 & 0 & 1 \end{bmatrix}$$

Relative to the axes $0xyz$ a vector $\mathbf{a}$ has components $(-1, 2, 1)$. Find the components of $\mathbf{a}$ relative to the axes $0x'y'z'$.

8.5 Parametric equations and vector-valued functions

We can represent a curve in two-dimensional space in terms of a parameter t in the form

$$x = x(t) \qquad y = y(t)$$

where $(x(t), y(t))$ are the rectangular coordinates of the point on the curve at the value t of some parameter, defined on a given range $a \le t \le b$ (which may be taken as $-\infty < t < \infty$ if not otherwise specified). Similarly, we can represent a curve in 3-space by the parametric equations

$$x = x(t) \quad y = y(t) \quad z = z(t)$$

We can express such parametric equations in the form of a vector

$$\mathbf{r} = x\mathbf{i} + y\mathbf{j} = x(t)\mathbf{i} + y(t)\mathbf{j} = \mathbf{r}(t)$$
$$\mathbf{r} = x\mathbf{i} + y\mathbf{j} + z\mathbf{k} = x(t)\mathbf{i} + y(t)\mathbf{j} + z(t)\mathbf{k} = \mathbf{r}(t)$$

Such expressions are called *vector-valued functions* of a real variable because they associate vectors with a real variable t. $x(t)$, $y(t)$, $z(t)$ are called the *component functions* of $\mathbf{r}(t)$. The range of allowed values of t is called the *domain* of $\mathbf{r}(t)$. The *graph* of $\mathbf{r}(t)$ is the curve in 2- or 3-space described as t varies. If $\mathbf{r}(t)$ is regarded as the tip of the vector from the origin to the point $\mathbf{r}(t)$ on the curve then $\mathbf{r}(t)$ is referred to as a *position vector*.

The vector function $\mathbf{r}(t) = x(t)\mathbf{i} + y(t)\mathbf{j} + z(t)\mathbf{k}$ is *differentiable* at $t = t_0$ if x, y, z are each differentiable at $t = t_0$. The curve traced by $\mathbf{r}(t)$ is said to be *smooth* if $d\mathbf{r}/dt$ is continuous and never zero. A curve made up of a finite number of smooth curves pieced together in a continuous fashion is called *piecewise smooth*. A piecewise smooth curve is said to be *simple* if every point corresponds to a unique value of t. If only the initial and final points coincide, i.e. $\mathbf{r}(a) = \mathbf{r}(b)$ and all other points correspond to unique values of t, then we say that the curve is *simple-closed*. We will have much to do with such simple-closed curves in Chapter 10.

1. What are the parametric equations of the curves described by the vectors
   (i) $t\mathbf{i} + t^2\mathbf{j} + t^3\mathbf{k}$ (ii) $e^t\mathbf{i} - 2t\mathbf{j} + (\cos t)\mathbf{k}$?
2. Sketch the graph of the curve represented by $\mathbf{r}(t) = (\cos t)\mathbf{i} + (\sin t)\mathbf{j} + t\mathbf{k}$.

8.6 Calculus of vector-valued functions

Since the basis vectors $\mathbf{i}$, $\mathbf{j}$, $\mathbf{k}$ are constant vectors, and the only dependence on t resides in the components of $\mathbf{r}(t)$, then it seems natural that whatever we do to $\mathbf{r}(t)$ as a function of t, we do to each of its components.

For example, by definition, for a vector-valued function in 2-space we have

$$\lim_{t \to a} \mathbf{r}(t) = \left(\lim_{t \to a} x(t)\right)\mathbf{i} + \left(\lim_{t \to a} y(t)\right)\mathbf{j}$$

A direct consequence of this is that we differentiate/integrate $\mathbf{r}(t)$ by differentiating/integrating each of its components separately:

$$\mathbf{r}'(t) = x'(t)\mathbf{i} + y'(t)\mathbf{j}$$

$$\int \mathbf{r}(t)\,dt = \left(\int x(t)\,dt\right)\mathbf{i} + \left(\int y(t)\,dt\right)\mathbf{j} + \mathbf{C}$$

(note that in this case the arbitrary integration constant is itself a vector)

$$\int_a^b \mathbf{r}(t)\,dt = \left(\int_a^b x(t)\,dt\right)\mathbf{i} + \left(\int_a^b y(t)\,dt\right)\mathbf{j}$$

The extensions to 3-space are obvious.

All the usual qualifications and terminology of ordinary calculus of a single variable (e.g. differentiability implies continuity) apply equally well to vector-valued functions if applied to their components. In particular, the derivative $\mathbf{r}'(t)$ is defined by

$$\mathbf{r}'(t) = \lim_{h \to 0}\left[\frac{\mathbf{r}(t+h) - \mathbf{r}(t)}{h}\right]$$

PROBLEM 7

Apply the above results to

$$\mathbf{r}(t) = t\mathbf{i} - (\sin t)\mathbf{j} + e^{-3t}\mathbf{k}$$

$$\lim_{t \to a} \mathbf{r}(t) = a\mathbf{i} - (2\sin a)\mathbf{j} + e^{-3a}\mathbf{k}$$

$$\mathbf{r}'(t) = \mathbf{i} - (2\cos t)\mathbf{j} - 3e^{-3t}\mathbf{k}$$

$$\int \mathbf{r}(t)\, dt = \left(\frac{t^2}{2} + A\right)\mathbf{i} + (2\cos t + B)\mathbf{j} + \left(-\frac{1}{3}e^{-3t} + C\right)\mathbf{k}$$

$$= \frac{t^2}{2}\mathbf{i} + 2(\cos t)\mathbf{j} - \frac{1}{3}e^{-3t}\mathbf{k} + A\mathbf{i} + B\mathbf{j} + C\mathbf{k}$$

Here the arbitrary integration constant is the vector $A\mathbf{i} + B\mathbf{j} + C\mathbf{k}$, where A, B, C are arbitrary constants.

An example of a definite integral is

$$\int_0^1 \mathbf{r}(t)\, dt = \left[\frac{t^2}{2}\right]_0^1 \mathbf{i} + [2\cos t]_0^1 \mathbf{j} - \left[\frac{1}{3}e^{-3t}\right]_0^1 \mathbf{k}$$

$$= \frac{1}{2}\mathbf{i} + 2(\cos 1 - 1)\mathbf{j} - \frac{1}{3}e^{-3}\mathbf{k}$$

The componentwise definition of vector-valued functions allows us to extend the usual rules of calculus to such functions. Thus, in either 2- or 3-space, if $\mathbf{r}(t)$, $\mathbf{r}_1(t)$, $\mathbf{r}_2(t)$ are appropriately well-behaved vector-valued functions, $\mathbf{c}$ is any constant vector and a, b and k are arbitrary constant scalars, then the rules of differentiation are:

$$\frac{d}{dt}[\mathbf{c}] = 0$$

$$\frac{d}{dt}[k\mathbf{r}(t)] = k\frac{d\mathbf{r}(t)}{dt}$$

$$\frac{d}{dt}[a\mathbf{r}_1(t) + b\mathbf{r}_2(t)] = a\frac{d\mathbf{r}_1(t)}{dt} + b\frac{d\mathbf{r}_2(t)}{dt}$$

and

$$\frac{d}{dt}[f(t)\mathbf{r}(t)] = f(t)\frac{d\mathbf{r}(t)}{dt} + \frac{d[f(t)]}{dt}\mathbf{r}(t)$$

where $f(t)$ is any differentiable scalar function.

The derivatives of the scalar and vector products are as follows:

$$\frac{d}{dt}[\mathbf{r}_1(t) \cdot \mathbf{r}_2(t)] = \mathbf{r}_1(t) \cdot \frac{d\mathbf{r}_2(t)}{dt} + \frac{d\mathbf{r}_1(t)}{dt} \cdot \mathbf{r}_2(t)$$

$$\frac{d}{dt}[\mathbf{r}_1(t) \times \mathbf{r}_2(t)] = \mathbf{r}_1(t) \times \frac{d\mathbf{r}_2(t)}{dt} + \frac{d\mathbf{r}_1(t)}{dt} \times \mathbf{r}_2(t)$$

Note that the order of the vectors has to be preserved in the case of the vector product.

The rules of integration can be summarized in the linearity condition

$$\int [a\mathbf{r}_1(t) + b\mathbf{r}_2(t)]\, dt = a\int \mathbf{r}_1(t)\, dt + b\int \mathbf{r}_2(t)\, dt$$

which also applies to definite integrals. These can be evaluated in the usual way: if

$$\int \mathbf{r}(t)\, dt = \mathbf{R}(t) + \mathbf{C}$$

where $\mathbf{C}$ is an arbitrary constant vector, then

$$\int_a^b \mathbf{r}(t)\, dt = \mathbf{R}(t)\Big|_a^b = \mathbf{R}(b) - \mathbf{R}(a)$$

EXERCISE ON 8.6

Given that $\mathbf{r}_1(t) = t\mathbf{i} + 2t^2\mathbf{j} + e^t\mathbf{k}$ and $\mathbf{r}_2(t) = (t^2 - 1)\mathbf{i} + 3(\cos t)\mathbf{j} - t\mathbf{k}$, evaluate

(i) $\mathbf{r}_1 \cdot \mathbf{r}_2$ (ii) $\mathbf{r}_2'$ (iii) $\mathbf{r}_1 + 2\mathbf{r}_2$ (iv) $\mathbf{r}_1 \times \mathbf{r}_2$

(v) $(\mathbf{r}_1 \cdot \mathbf{r}_2)'$ (vi) $(\mathbf{r}_1 + 2\mathbf{r}_2)'$ (vii) $(\mathbf{r}_1 \times \mathbf{r}_2)'$ (viii) $\int_0^1 \mathbf{r}_1 \cdot \mathbf{r}_2\, dt$

(ix) $\int_{-1}^1 (\mathbf{r}_1 + 2\mathbf{r}_2)\, dt$ (x) $|\mathbf{r}_1|'$

8.7 Curves and the tangent vector

Here we look more closely at the idea of a vector-valued function viewed as a curve. Suppose

$$\mathbf{r}(t) = x(t)\mathbf{i} + y(t)\mathbf{j} + z(t)\mathbf{k}$$

is differentiable on some interval I, and is continuous at its endpoints. For every $t \in I$ the tip of the radius vector $\mathbf{r}(t)$ is the point $P(x(t), y(t), z(t))$, which traces out some path C as t varies over I. C is called a *differentiable curve parametrized by t*. C is said to be *oriented* in the direction in which t increases on I.

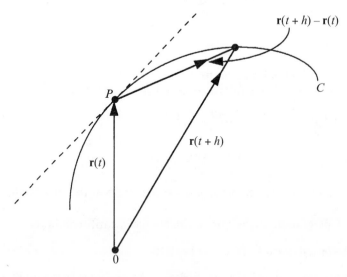

Fig. 8.1

Given a differentiable curve we may want to find the tangent at any point. We can approach this geometrically by considering Fig. 8.1. The position vectors $\mathbf{r}(t)$,

$\mathbf{r}(t+h)$ specify two points on the curve, with $\mathbf{r}(t)$ at the point P. The vector $\mathbf{r}(t+h) - \mathbf{r}(t)$ is a chord of the curve, through P. Now form the vector

$$\frac{\mathbf{r}(t+h) - \mathbf{r}(t)}{h}$$

Clearly, as h approaches zero, this vector points in the direction of the tangent. The result is, of course, the *derivative* of $\mathbf{r}$:

$$\mathbf{r}'(t) = \lim_{h \to 0} \left[\frac{\mathbf{r}(t+h) - \mathbf{r}(t)}{h} \right]$$

We take this to be the tangent to the curve. It points in the direction of increasing t along C. So we define the *tangent vector to the differentiable curve C at the point* $P(x(t), y(t), z(t))$ to be the vector

$$\mathbf{r}'(t) = x'(t)\mathbf{i} + y'(t)\mathbf{j} + z'(t)\mathbf{k}$$

provided it is non-zero.

Suppose the curve

$$C : \mathbf{r}(t) = x(t)\mathbf{i} + y(t)\mathbf{j} + z(t)\mathbf{k}$$

is twice differentiable and its derivative is never zero. Then at each point of the curve we can define a *unit tangent vector*

$$\mathbf{T}(t) = \frac{\mathbf{r}'(t)}{|\mathbf{r}'(t)|}$$

where $|\mathbf{a}|$ denotes the norm or length of the vector $\mathbf{a}$. $\mathbf{T}(t)$ points in the direction of $\mathbf{r}'(t)$.

PROBLEM 8

> Prove that the vector $\mathbf{T}'(t)$ is perpendicular to $\mathbf{T}(t)$.

Since $\mathbf{T}(t)$ is a unit vector we have

$$\mathbf{T}(t) . \mathbf{T}(t) = 1$$

Differentiating gives

$$\mathbf{T}'(t) . \mathbf{T}(t) + \mathbf{T}(t) . \mathbf{T}'(t) = 0$$

or

$$\mathbf{T}'(t) . \mathbf{T}(t) = 0$$

i.e. $\mathbf{T}'(t)$ is perpendicular to $\mathbf{T}(t)$.

This enables us to give a geometrical interpretation to $\mathbf{T}'(t)$ – it measures the rate of change of direction of $\mathbf{T}(t)$. This is so because the norm of $\mathbf{T}(t)$ is always constant (unity) and so it can only change in direction – which is therefore what $\mathbf{T}'(t)$ measures.

If $\mathbf{T}'(t) \neq 0$, then we can construct the *principal normal vector*

$$\mathbf{N}(t) = \frac{\mathbf{T}'(t)}{|\mathbf{T}'(t)|}$$

which is the unit vector in the direction of $\mathbf{T}'(t)$. The plane defined by the unit tangent vector and the principal normal vector is called the *osculating plane* – named so because it 'kisses' the curve C. It is the plane of greatest contact with the curve at a given point.

EXERCISE ON 8.7

For the curves

(i) $\mathbf{r}_1(t) = (\cos t)\mathbf{i} + (\sin t)\mathbf{j} + t\mathbf{k}$ (ii) $\mathbf{r}_2(t) = \dfrac{t^3}{3}\mathbf{i} + \dfrac{t}{\sqrt{2}}\mathbf{j} + t\mathbf{k}$

obtain the unit tangent vector and the principal normal vector. In each case verify that these are perpendicular to each other.

8.8 Arc length

Given a differentiable curve, one parameter of obvious importance is the arc length – the distance measured along the curve from some fixed origin. Before giving a precise definition of arc length, let us think about how we might try to evaluate it directly.

Suppose a curve C in 3-space is defined by continuously differentiable functions $x = x(t)$, $y = y(t)$, $z = z(t)$ parametrized by a parameter $t \in [a, b]$ in such a way that as t varies from a to b, C is described once only. We can approximate the curve by a series of line segments joining points P_0, P_1, ... , P_n on the curve defined by some sequence of values of t:

$$a < t_0 < t_1 < t_2 < ... < t_{i-1} < t_i < ... < t_{n-1} < t_n < b$$

Here P_i denotes the point $P(x(t_i), y(t_i), z(t_i))$. These points, connected by the line segments, form a polygonal path approximating C. Different choices of sets of values of t yield different polygonal approximations to C. For each such approximation we can easily calculate the length of each line segment, and totalling up the results gives an approximation to the length of C. Let a typical polygonal path be denoted by γ and its length by $s(\gamma)$. Clearly, for any such path we will have $s(\gamma) \leq s(C)$, the length of C. Furthermore, by choosing as large a number of points as we wish we would expect to be able to approximate C as closely as we wish. In particular, there should be a path γ whose length approximates that of C as closely as we wish. This leads to the definition:

● *Arc length* ——————————————————————————————

The *arc length* of C, $s(C)$, is the least upper bound of the set of all lengths of polygonal paths constructed on C.

Using this definition, and the idea of the integral as the limit of a sum, it can be shown that the arc length of the curve in 2-space C: $x = x(t)$, $y = y(t)$, $t \in [a, b]$, is

$$s(C) = \int_a^b \sqrt{[x'(t)]^2 + [y'(t)]^2}\, dt$$

Similarly, for the curve in 3-space $x = x(t)$, $y = y(t)$, $z = z(t)$, $t \in [a, b]$, the arc length is

$$s(C) = \int_a^b \sqrt{[x'(t)]^2 + [y'(t)]^2 + [z'(t)]^2} \, dt$$

These results can be written in the vector form

$$s(C) = \int_a^b |\mathbf{r}'(t)| \, dt$$

PROBLEM 9

Find the length of the curve

$C : \mathbf{r}(t) = 3(\cos t)\mathbf{i} + 3(\sin t)\mathbf{j} + 4t\mathbf{k}$

from $t = 0$ to $t = 1$.

We have

$$s(C) = \int_0^1 \sqrt{9 \sin^2 t + 9 \cos^2 t + 4^2} \, dt = 5 \int_0^1 dt = 5$$

If $\mathbf{r}(t)$ is continuously differentiable in the interval $[a, b]$ and $\int_a^{t_1} |\mathbf{r}'(t)| \, dt$ exists for every t_1 in $[a, b]$ then we say that the curve $\mathbf{r}(t)$ is *rectifiable* on $[a, b]$.

The arc length s is obviously a useful parameter for a curve. In particular, when s is used as a parameter an important relationship between position vectors and tangent vectors is revealed. Thus, by definition:

$$\frac{ds}{dt} = |\mathbf{r}'(t)|$$

for *any* parameter t. So, using s as a parameter, the unit tangent vector can be written

$$\mathbf{T}(s) = \frac{d\mathbf{r}/ds}{|d\mathbf{r}/ds|} = \frac{d\mathbf{r}}{ds}$$

on using

$$\left| \frac{d\mathbf{r}}{ds} \right| = \left| \frac{d\mathbf{r}/dt}{ds/dt} \right| = \left| \frac{\mathbf{r}'(t)}{ds/dt} \right| = 1$$

EXERCISE ON 8.8

Find the length of the following arcs over the specified interval:

(i) $\frac{2}{3}t^3\mathbf{i} + (1 + t^{9/2})\mathbf{j} + (1 - t^{9/2})\mathbf{k}$ $t \in [0, 2]$

(ii) $e^t(\cos 2t)\mathbf{i} + e^t(\sin 2t)\mathbf{j} + e^t\mathbf{k}$ $t \in [1, 4]$

8.9   Curvature and torsion

The curvature of a curve measures how sharply it bends at any given point. So it depends on how quickly the tangent vector **T** changes as we move along the curve. This is measured by the rate of change of **T** with arc length s and leads to the following definition.

● *Curvature*

If C is a smooth curve in 2- or 3-space, and if the arc length of C is represented by the parameter s, then the *curvature* of C, denoted by the Greek letter 'kappa', $\kappa(s)$, is defined to be the positive quantity

$$\kappa(s) = \left| \frac{d\mathbf{T}}{ds} \right| \tag{8.6}$$

In general, the curvature will change as we travel along the curve. However, a notable exception to this is the circle.

PROBLEM 10

Determine the curvature of a circle with radius a.

The circle of radius a, centred at the origin, can be parametrized in terms of arc length as

$$\mathbf{r}(s) = a \cos\left(\frac{s}{a}\right)\mathbf{i} + a \sin\left(\frac{s}{a}\right)\mathbf{j} \qquad s \in [0, 2\pi a]$$

So

$$\frac{d\mathbf{T}}{ds} = \frac{d^2\mathbf{r}}{ds^2} = -\frac{1}{a}\cos\left(\frac{s}{a}\right)\mathbf{i} - \frac{1}{a}\sin\left(\frac{s}{a}\right)\mathbf{j}$$

Hence

$$\kappa = \left| \frac{d\mathbf{T}}{ds} \right| = \sqrt{\left[-\frac{1}{a}\cos\left(\frac{s}{a}\right)\right]^2 + \left[-\frac{1}{a}\sin\left(\frac{s}{a}\right)\right]^2} = \frac{1}{a}$$

So the circle has constant curvature $1/a$, the inverse of the radius.

For arbitrary parametrizations of a curve, the following alternative forms are more convenient:

$$\kappa = \kappa(t) = \frac{|\mathbf{T}'(t)|}{|\mathbf{r}'(t)|} = \frac{|\mathbf{r}'(t) \times \mathbf{r}''(t)|}{|\mathbf{r}'(t)|^3} \tag{8.7}$$

The first form follows from

$$\kappa = \left| \frac{d\mathbf{T}}{ds} \right| = \left| \frac{d\mathbf{T}/dt}{ds/dt} \right| = \left| \frac{d\mathbf{T}/dt}{dr/dt} \right| = \frac{|\mathbf{T}'(t)|}{|\mathbf{r}'(t)|}$$

For the second form note that since $\mathbf{r}'(t) = |\mathbf{r}'(t)|\,\mathbf{T}(t)$ then

$$\mathbf{r}''(t) = |\mathbf{r}'(t)|'\,\mathbf{T}(t) + |\mathbf{r}'(t)|\,\mathbf{T}'(t)$$

But by definition

$$\mathbf{T}'(t) = |\mathbf{T}'(t)|\,\mathbf{N}(t) = \kappa\,|\mathbf{r}'(t)|\,\mathbf{N}(t)$$

Using this in $\mathbf{r}''(t)$, we have

$$\mathbf{r}''(t) = |\mathbf{r}'(t)|'\,\mathbf{T}(t) + \kappa\,|\mathbf{r}'(t)|^2\,\mathbf{N}(t)$$

Hence

$$\mathbf{r}'(t) \times \mathbf{r}''(t) = |\mathbf{r}'(t)|\,|\mathbf{r}'(t)|'(\mathbf{T}(t) \times \mathbf{T}(t))$$
$$+ \kappa\,|\mathbf{r}'(t)|^3\,(\mathbf{T}(t) \times \mathbf{N}(t))$$
$$= \kappa\,|\mathbf{r}'(t)|^3\,(\mathbf{T}(t) \times \mathbf{N}(t))$$

Since $\mathbf{T}(t) \times \mathbf{N}(t)$ is a unit vector, it follows that

$$|\mathbf{r}'(t) \times \mathbf{r}''(t)| = \kappa\,|\mathbf{r}'(t)|^3$$

proving the result (8.7).

Suppose the curve C has a non-zero curvature κ at a point P. Then at P there will be a circle sharing a common tangent at P, whose radius is $1/\kappa = \rho$, say. This is called the *radius of curvature*. The circle is called the *circle of curvature* or *osculating circle* at P. The circle of curvature is the circle which 'best' approximates the curve C near P. The centre of the circle of curvature is called the *centre of curvature*.

There is an important relationship between $\mathbf{T}$, $\mathbf{N}$ and κ, derivable from the definition of $\mathbf{N}(t)$. We have

$$\mathbf{N}(s) = \frac{d\mathbf{T}/ds}{|\,d\mathbf{T}/ds\,|} = \frac{1}{\kappa}\frac{d\mathbf{T}}{ds}$$

so

$$\frac{d\mathbf{T}}{ds} = \kappa\mathbf{N}$$

If we think of $\mathbf{T}$ and $\mathbf{N}$ at a particular point on a curve $\mathbf{r}(t)$, we see that they define a plane through which the curve passes at that point. For many purposes it is useful to define a further vector, the *unit binormal vector*, at this point by

$$\mathbf{B} = \mathbf{T} \times \mathbf{N}$$

Then the three unit vectors $\mathbf{T}$, $\mathbf{N}$, $\mathbf{B}$ form an orthonormal right-handed triad at the given point (see Fig. 8.2).

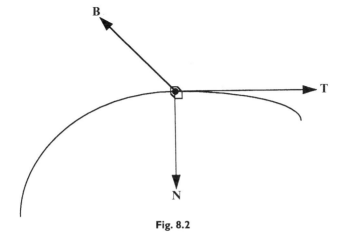

Fig. 8.2

PROBLEM 11

Prove that d**B**/d*s* is parallel or anti-parallel to **N**.

We have

$$\frac{d\mathbf{B}}{ds} = \frac{d}{ds}(\mathbf{T} \times \mathbf{N}) = \frac{d\mathbf{T}}{ds} \times \mathbf{N} + \mathbf{T} \times \frac{d\mathbf{N}}{ds}$$

$$= \kappa\mathbf{N} \times \mathbf{N} + \mathbf{T} \times \frac{d\mathbf{N}}{ds} = \mathbf{T} \times \frac{d\mathbf{N}}{ds}$$

Since d**B**/d*s* is normal to **B** (why?), it must lie in the plane of **N** and **T**. However, **T** × d**N**/d*s* is clearly normal to **T** and so d**B**/d*s* must be parallel (or anti-parallel) to **N**. We may therefore write

$$\frac{d\mathbf{B}}{ds} = -\tau\mathbf{N}$$

where τ is some function of *s* (the sign is conventional). τ is called the *torsion* of the curve. It is a measure of the rate at which the direction of the binormal changes with *s*. Geometrically this is the rate at which the curve is leaving the osculating plane.

PROBLEM 12

Show that

$$\frac{d\mathbf{N}}{ds} = -\kappa\mathbf{T} + \tau\mathbf{B}$$

Since **T**, **N**, **B** form an orthonormal set we have **N** = **B** × **T**. So

$$\frac{d\mathbf{N}}{ds} = \frac{d}{ds}(\mathbf{B} \times \mathbf{T}) = \frac{d\mathbf{B}}{ds} \times \mathbf{T} + \mathbf{B} \times \frac{d\mathbf{T}}{ds}$$

$$= -\tau\mathbf{N} \times \mathbf{T} + \mathbf{B} \times \kappa\mathbf{N}$$

$$= -\tau(-\mathbf{B}) + \kappa(-\mathbf{T})$$

$$= -\kappa\mathbf{T} + \tau\mathbf{B}$$

The three equations

$$\frac{d\mathbf{T}}{ds} = \kappa\mathbf{N}$$

$$\frac{d\mathbf{B}}{ds} = -\tau\mathbf{N}$$

$$\frac{d\mathbf{N}}{ds} = -\kappa\mathbf{T} + \tau\mathbf{B}$$

are key results in the theory of the differential geometry of curves. They are called the *Serret–Frenet formulae*. They were originally developed by the French geometer Frederic Frenet (1816–1900) in 1847 and independently rediscovered by his fellow countryman Alfred Serret (1819–85) in 1850. The triad of vectors **T, N, B** is referred to as the *Frenet frame*, or simply the **TNB** frame.

EXERCISES ON 8.9

1. Obtain the vectors **T, N, B** for the *circular helix* **r** = a (cos t) **i** + a (sin t) **j** + bt **k** and show that $\kappa = |a| / (a^2 + b^2)$ and $\tau = b / (a^2 + b^2)$.
2. Show that for a plane curve in the xy-plane, if the unit tangent in the direction of s increasing makes an angle ψ with the positive x-axis, then the radius of curvature is given by

$$\rho = \left| \frac{ds}{d\psi} \right|$$

FURTHER EXERCISES

1. Show that the following matrices, **T**, describe a rotation of a rectangular coordinate system, $0xyz$, about the origin. In each case describe the form of the rotation in words (cf. Exercises 8.2).

(i) $T = \begin{bmatrix} \dfrac{1}{2} & \dfrac{\sqrt{3}}{2} & 0 \\ -\dfrac{\sqrt{3}}{2} & \dfrac{1}{2} & 0 \\ 0 & 0 & 1 \end{bmatrix}$
(ii) $T = \begin{bmatrix} 1 & 0 & 0 \\ 0 & \dfrac{\sqrt{3}}{2} & \dfrac{1}{2} \\ 0 & -\dfrac{1}{2} & \dfrac{\sqrt{3}}{2} \end{bmatrix}$

(iii) $T = \begin{bmatrix} \dfrac{1}{\sqrt{2}} & 0 & \dfrac{1}{\sqrt{2}} \\ 0 & 1 & 0 \\ -\dfrac{1}{\sqrt{2}} & 0 & \dfrac{1}{\sqrt{2}} \end{bmatrix}$

2. Show that the following matrices each represent a rotation of a rectangular coordinate system, $0xyz$:

(i) $\begin{bmatrix} 1 & 0 & 0 \\ 0 & 0 & 1 \\ 0 & 1 & 0 \end{bmatrix}$
(ii) $\begin{bmatrix} \dfrac{1}{2} & \dfrac{\sqrt{3}}{2} & 0 \\ 0 & 0 & 1 \\ -\dfrac{\sqrt{3}}{2} & \dfrac{1}{2} & 0 \end{bmatrix}$
(iii) $\dfrac{1}{7}\begin{bmatrix} -3 & -6 & 2 \\ -2 & 3 & -6 \\ 6 & -2 & -3 \end{bmatrix}$

3. Write down each of the following using the summation convention:

(i) $a_1 x_1 x_2 + a_2 x_2^2 + \ldots + a_N x_N x_2$

(ii) $\dfrac{dy_k}{dt} = \dfrac{\partial y_k}{\partial x_1}\dfrac{dx_1}{dt} + \dfrac{\partial y_k}{\partial x_2}\dfrac{dx_2}{dt} + \ldots + \dfrac{\partial y_N}{\partial x_N}\dfrac{dx_N}{dt}$

4. If x_i, $i = 1, 2, 3$, are the components of a vector, prove that dx_i is also a vector. Deduce that the 'velocity' $v_i = dx_i/dt$ is also a vector. Show that acceleration $d^2x_i/dt^2 = a_i$ is also a vector. Given that force is a vector, use Newton's second law, $\mathbf{F} = m\mathbf{a}$, to determine the nature of the mass m. Show that the following are scalars:

(i) $\mathbf{F} \cdot \mathbf{r}$ (ii) $\mathbf{F} \cdot \mathbf{v}$ (iii) $\mathbf{a} \cdot \mathbf{v}$ (iv) $\mathbf{r} \cdot \mathbf{v}$

5. Find (a) $\mathbf{r}'$, (b) $\mathbf{r}''(t)$, (c) $|\mathbf{r}'(t)|$ and (d) $|\mathbf{r}''(t)|$ for each of the following:

(i) $\mathbf{r}(t) = (\cos t)\mathbf{i} + (\sin t)\mathbf{j} + t\mathbf{k}$ (ii) $\mathbf{r}(t) = e^t\mathbf{i} + 2e^{-2t}\mathbf{j} - (\cos t)\mathbf{k}$

(iii) $\mathbf{r}(t) = t\mathbf{i} - 2t^2\mathbf{j} + 3t\mathbf{k}$ (iv) $\mathbf{r}(t) = (\cos t)\mathbf{i} - (\sin t)\mathbf{j} + (\cos 2t)\mathbf{k}$

6. Given that $\mathbf{r}_1(t) = t\mathbf{i} + t^2\mathbf{j} - 3t\mathbf{k}$ and $\mathbf{r}_2(t) = t^2\mathbf{i} - 2\mathbf{j} + t\mathbf{k}$, evaluate:

(i) $(\mathbf{r}_1 \cdot \mathbf{r}_2)'$ (ii) $(2\mathbf{r}_1 - \mathbf{r}_2)''$ (iii) $\mathbf{r}_1 \times (2\mathbf{r}_1 + \mathbf{r}_2)'$

(iv) $(\mathbf{r}_1 \times \mathbf{r}_2)'$ (v) $\int \mathbf{r}_1 \cdot \mathbf{r}_2 \, dt$ (vi) $\int \mathbf{r}_1 \times \mathbf{r}_2 \, dt$

(vii) $\int_{-1}^{0} \mathbf{r}_1 \cdot (\mathbf{r}_1 + 2\mathbf{r}_2) \, dt$ (viii) $\int_{0}^{2} \mathbf{r}_1 \times \mathbf{r}_2 \, dt$

(ix) $|\mathbf{r}_1 \times \mathbf{r}_2|'$ (x) $\left(\dfrac{\mathbf{r}_1}{|\mathbf{r}_2|}\right)'$

7. If $\mathbf{a}(t)$, $\mathbf{b}(t)$ are differentiable functions of a scalar t, prove from first principles that

(i) $(\mathbf{a} \cdot \mathbf{b})' = \mathbf{a} \cdot \mathbf{b}' + \mathbf{a}' \cdot \mathbf{b}$ (ii) $(\mathbf{a} \times \mathbf{b})' = \mathbf{a}' \times \mathbf{b} + \mathbf{a} \times \mathbf{b}'$

8. Evaluate $(\mathbf{v} \cdot \mathbf{v}' \times \mathbf{v}'')'$.

9. Prove that

(i) $\mathbf{a} \times \mathbf{b}'' - \mathbf{a}'' \times \mathbf{b} = (\mathbf{a} \times \mathbf{b}' - \mathbf{a}' \times \mathbf{b})'$ (ii) $\mathbf{a} \cdot \mathbf{a}' = |\mathbf{a}| \, |\mathbf{a}|'$

10. Find the first and second derivatives of the following vector functions:

(i) $e^{\cos t}\mathbf{i} + e^{\sin t}\mathbf{j} + t\mathbf{k}$ (ii) $((\cos t)\mathbf{i} - (\sin t)\mathbf{j}) \cdot (e^t\mathbf{j} + t\mathbf{k})$

(iii) $(\mathbf{a} + t\mathbf{b}) \times (\mathbf{c} + t\mathbf{d})$ (iv) $(\mathbf{a} + t\mathbf{b} + t^2\mathbf{c}) \cdot (\mathbf{a} + t\mathbf{b})$

11. Show that in the following cases $\mathbf{r}(t)$ and $\mathbf{r}''(t)$ are parallel:

(i) $\mathbf{r}(t) = (\sin t)\mathbf{i} + (\cos t)\mathbf{j}$ (ii) $e^{kt}\mathbf{i} + e^{-kt}\mathbf{j}$

12. Find the point P on $\mathbf{r}(t) = (1 - 2t)\mathbf{i} + t^2\mathbf{j} + 2e^{2(t-1)}\mathbf{k}$ at which the tangent vector $\mathbf{r}'(t)$ is parallel to the radius vector $\mathbf{r}(t)$.

13. Show that $\mathbf{r} = e^{-2t}(\mathbf{C}_1 \cos 2t + \mathbf{C}_2 \sin 2t)$, where $\mathbf{C}_1$ and $\mathbf{C}_2$ are constant vectors, is a solution of the differential equation

$$\frac{d^2\mathbf{r}}{dt^2} + 4\frac{d\mathbf{r}}{dt} + 8\mathbf{r} = 0$$

14. If $\mathbf{C}_1$ and $\mathbf{C}_2$ are constant vectors and λ is a constant scalar, show that $\mathbf{H} = e^{-\lambda x}(\mathbf{C}_1 \sin \lambda y + \mathbf{C}_2 \cos \lambda y)$ satisfies the partial differential equation

$$\frac{\partial^2 \mathbf{H}}{\partial x^2} + \frac{\partial^2 \mathbf{H}}{\partial y^2} = 0$$

15. Prove that $\mathbf{A} = \mathbf{p}_0 e^{i\omega(t - r/c)}/r$, where $\mathbf{p}_0$ is a constant vector, ω and c are constant scalars and $i = \sqrt{-1}$, satisfies the equation

$$\frac{\partial^2 \mathbf{A}}{\partial r^2} + \frac{2}{r} \frac{\partial \mathbf{A}}{\partial r} = \frac{1}{c^2} \frac{\partial^2 \mathbf{A}}{\partial t^2}$$

This result is of importance in *electromagnetic theory*.

16. If $\mathbf{F} = (x^2 y - 2z)\mathbf{i} + xy^2 z\mathbf{j} + x \cos(xz)\mathbf{k}$ find all first- and second-order partial derivatives of $\mathbf{F}$.

17. For the curve $\mathbf{r} = 3(\cos t)\mathbf{i} + 3(\sin t)\mathbf{j} + 4t\mathbf{k}$ find: (i) the unit tangent $\mathbf{T}$, (ii) the principal normal $\mathbf{N}$, curvature, κ, and the radius of curvature ρ, (iii) the binormal $\mathbf{B}$, and the torsion τ.

18. If s denotes arc length, prove that the radius of curvature of the curve with parametric equations $x = x(s)$, $y = y(s)$, $z = z(s)$ is given by

$$\rho = \left[\left(\frac{d^2 x}{ds^2} \right)^2 + \left(\frac{d^2 y}{ds^2} \right)^2 + \left(\frac{d^2 z}{ds^2} \right)^2 \right]^{-1/2}$$

19. Show that, if s is arc length, then

$$\frac{d\mathbf{r}}{ds} \cdot \frac{d^2\mathbf{r}}{ds^2} \times \frac{d^3\mathbf{r}}{ds^3} = \frac{\tau}{\rho^2}$$

20. Find (a) the unit tangent $\mathbf{T}$ and the curvature κ, (b) the principal normal $\mathbf{N}$, the binormal $\mathbf{B}$, and the torsion τ for the following space curves:

(i) $x = t$, $y = t^2$, $z = \frac{2}{3}t^3$ (twisted cubic)

(ii) $x = e^t$, $y = e^{-t}$, $z = \sqrt{2}t$

(iii) $x = t - \dfrac{t^3}{3}$, $y = t^2$, $z = t + \dfrac{t^3}{3}$

21. In the following find the unit tangent vector, the principal normal vector, and an equation in x, y, z for the osculating plane at the point on the curve corresponding to the indicated value of t:

(i) $\mathbf{r}(t) = \mathbf{i} + 2t\mathbf{j} + t^2\mathbf{k}$	$t = 1$
(ii) $\mathbf{r}(t) = (\cos 2t)\mathbf{i} + (\sin 2t)\mathbf{j} + t\mathbf{k}$	$t = \pi/4$
(iii) $\mathbf{r}(t) = e^t(\sin t)\mathbf{i} + e^t(\cos t)\mathbf{j} + e^t\mathbf{k}$	$t = 0$

22. Essay and discussion topics:
    (i) Rotation of axes and the definition of a Cartesian vector.
    (ii) Scalar and vector quantities in mechanics.
    (iii) The calculus of vector-valued functions of a real variable.
    (iv) Parametrization of a curve by arc length.
    (v) Applications of curvature in mechanics.
    (vi) The Serret–Frenet formulae and their applications.

9 • Vector Differential Operators

9.1 Directional derivatives and the gradient

Suppose a function $f(x, y)$ is differentiable and defined on a differentiable curve $x = g(t)$, $y = h(t)$. Then we know from Section 4.5 that the rate of change of $f(x, y)$ with respect to the parameter t as we move along the curve is given by the chain rule

$$\frac{df}{dt} = \frac{\partial f}{\partial x}\frac{dx}{dt} + \frac{\partial f}{\partial y}\frac{dy}{dt}$$

An obvious question concerns the rate of change of f in any given direction, defined by a unit vector $\mathbf{u}$. We can approach this by considering the case when the curve in question is a straight line, passing through a point $P_0(x_0, y_0)$, in the direction of the unit vector $\mathbf{u} = u_1\mathbf{i} + u_2\mathbf{j}$. Then any point on the line is given by

$$x = x_0 + su_1 \qquad y = y_0 + su_2 \tag{9.1}$$

where s is some parameter. It is thus reasonable to define the *directional derivative* of f at $P_0(x_0, y_0)$ in the direction of the unit vector $\mathbf{u} = u_1\mathbf{i} + u_2\mathbf{j}$ by the limit

$$\left(\frac{df}{ds}\right)_{\mathbf{u}, P_0} = \lim_{s \to 0} \frac{f(x_0 + su_1, y_0 + su_2) - f(x_0, y_0)}{s}$$

provided it exists. We denote this directional derivative by

$$(D_{\mathbf{u}}f)_{P_0}$$

and refer to it as the *derivative of f at P_0, in the direction of $\mathbf{u}$.*

PROBLEM I

Find the derivative of

$$f(x, y) = x^2 + y^2$$

at $P_0(1, 1)$ in the direction $\mathbf{u} = \frac{1}{2}\mathbf{i} + \frac{\sqrt{3}}{2}\mathbf{j}$.

In this case the 'curve' is

$$x = 1 + \frac{1}{2}s \qquad y = 1 + \frac{\sqrt{3}}{2}s$$

So

$$\left(\frac{df}{ds}\right)_{\mathbf{u},P_0} = \lim_{s \to 0}\left(\frac{f\left(1 + \frac{1}{2}s, 1 + \frac{\sqrt{3}}{2}s\right) - f(1, 1)}{s}\right)$$

$$= \lim_{s \to 0}\left(\frac{\left(1 + \frac{1}{2}s\right)^2 + \left(1 + \frac{\sqrt{3}}{2}s\right)^2 - 2}{s}\right)$$

$$= 1 + \sqrt{3}$$

i.e. the rate of change of $f(x, y) = x^2 + y^2$ at $P_0(1, 1)$ in the direction $\mathbf{u} = \frac{1}{2}\mathbf{i} + \frac{\sqrt{3}}{2}\mathbf{j}$ is $1 + \sqrt{3}$.

Geometrically, the directional derivative is simply a measure of the steepness of the hill defined by $z = f(x, y)$ at the point $P_0(x_0, y_0)$, if we head in the direction of the vector $\mathbf{u}$.

We can obtain a simple formula for the directional derivative by using the chain rule. We have

$$\left(\frac{df}{ds}\right)_{\mathbf{u},P_0} = \left(\frac{\partial f}{\partial x}\right)_{P_0}\frac{dx}{ds} + \left(\frac{\partial f}{\partial y}\right)_{P_0}\frac{dy}{ds}$$

$$= \left(\frac{\partial f}{\partial x}\right)_{P_0}u_1 + \left(\frac{\partial f}{\partial y}\right)_{P_0}u_2$$

using the parametrization (9.1). By inspection, we notice that this could be written

$$\left(\frac{\partial f}{\partial s}\right)_{\mathbf{u},P_0} = \left(\left(\frac{\partial f}{\partial x}\right)_{P_0}\mathbf{i} + \left(\frac{\partial f}{\partial y}\right)_{P_0}\mathbf{j}\right) \cdot (u_1\mathbf{i} + u_2\mathbf{j})$$

which *looks like* the scalar product of two vectors. But before you get too excited, a question:

PROBLEM 2

The notation above is *suggestive* of a scalar product, but what condition must hold in order that it *is* a scalar product?

Well, we know that $\mathbf{u} = u_1\mathbf{i} + u_2\mathbf{j}$ is a vector – but we must also have that the object $\partial f/\partial x\,\mathbf{i} + \partial f/\partial y\,\mathbf{j}$ is a vector. This is not obvious – we must check how it transforms (*see* Section 8.4), which we will do later. Just for now assume that it is a vector. It is in fact a very important object, called the *gradient vector* (or *gradient*) of $f(x, y)$, and is denoted by

$$\nabla f \equiv \frac{\partial f}{\partial x}\mathbf{i} + \frac{\partial f}{\partial y}\mathbf{j} \equiv \operatorname{grad} f$$

This is read as 'grad f' or 'del f', since ∇ is read as 'del'. ∇ is also called *nabla*, which is apparently the Assyrian name for a harp!

So, with this definition we can write the directional derivative in the plane as

$$(D_u f)_{P_0} \equiv \left(\frac{df}{ds}\right)_{u,P_0} = (\nabla f)_{P_0} \cdot u \qquad (9.2)$$

i.e. as the scalar product of the gradient of f at P_0 and u. You will notice that in this form the coordinates (x, y) have vanished from view. This is, of course, the merit of vector notation – formulae are independent of the coordinate system. But, as noted above, this is only the case if the objects involved are truly vectors. And again, as we noted above, it is not obvious that ∇f is a vector – it is something we must prove. This idea will recur frequently – the need to ensure that novel objects we encounter are in fact of the type that we claim, in this case that ∇f is a vector. It is *not* enough to note that it is written in the form of a linear combination of i and j – that alone does not guarantee that it is a vector, because the coefficients of the combination vary with x and y and the total object ∇f does not necessarily transform like a vector.

PROBLEM 3

Repeat Problem 1 using the definition (9.2).

The gradient is

$$\nabla f = 2x i + 2y j$$

So at $P_0(1, 1)$:

$$(\nabla f)_{P_0} = 2i + 2j$$

The directional derivative is thus

$$\left(\frac{df}{ds}\right)_{u,P_0} = (2i + 2j) \cdot \left(\frac{1}{2}i + \frac{\sqrt{3}}{2}j\right) = 1 + \sqrt{3}$$

as before.

The scalar product form of the directional derivative yields some interesting properties of the gradient. Thus, let θ be the angle between the gradient ∇f and u. Then

$$\left(\frac{df}{ds}\right)_{u,P_0} = |\nabla f| \, |u| \cos \theta$$

$$= |\nabla f| \cos \theta$$

Clearly, this is a maximum when $\theta = 0$, i.e. when u is in the same direction as ∇f. Turning this round, we see that the gradient ∇f is the rate of change of f in the direction in which f changes the fastest. Also, when $\theta = \pi/2$, i.e. u is perpendicular to ∇f, then f does not change at all.

These facts highlight a very important geometrical property of the gradient which we will discuss in the next section. But first let us extend these ideas to three dimensions. This is fairly straightforward; we simply tag the extra dimension onto the two-dimensional definitions. Thus, the gradient of a function $f(x, y, z)$ on $\mathbb{R}^3$ is defined by

$$\nabla f = \frac{\partial f}{\partial x}\mathbf{i} + \frac{\partial f}{\partial y}\mathbf{j} + \frac{\partial f}{\partial z}\mathbf{k}$$

and the directional derivative is defined by

$$D_{\mathbf{u}}f = \nabla f \cdot \mathbf{u} = \frac{\partial f}{\partial x}u_1 + \frac{\partial f}{\partial y}u_2 + \frac{\partial f}{\partial z}u_3$$

This may still be written in the form

$$\nabla f \cdot \mathbf{u} = |\nabla f| \cos \theta$$

and so the properties obtained above still hold. That is, in three dimensions f increases most rapidly in the direction of ∇f, decreases most rapidly in the direction $-\nabla f$, and does not change at all in directions orthogonal to ∇f.

PROBLEM 4

Find the derivative of $f(x, y, z) = x^3 + 3xy^2 - 2z^2$ at the point $(-1, 0, 1)$ in the direction of $\mathbf{a} = 3\mathbf{i} + 2\mathbf{j} - \mathbf{k}$. In what directions does f increase and decrease most rapidly and what are the rates of change in these directions? In what direction does f not change at all?

A unit vector in the direction of $\mathbf{a}$ is

$$\mathbf{u} = \frac{1}{\sqrt{14}}(3\mathbf{i} + 2\mathbf{j} - \mathbf{k})$$

$$\nabla f = (3x^2 + 3y^2)\mathbf{i} + 6xy\mathbf{j} - 4z\mathbf{k}$$

and so at $(-1, 0, 1)$ we have

$$\nabla f = 3\mathbf{i} - 4\mathbf{k}$$

So the derivative in the direction of $\mathbf{u}$ is

$$\nabla f \cdot \mathbf{u} = \frac{13}{\sqrt{14}}$$

f increases most rapidly in the direction of $3\mathbf{i} - 4\mathbf{k}$ and decreases most rapidly in the direction $-3\mathbf{i} + 4\mathbf{k}$. The rates of change in these directions are

$$\pm |\nabla f| = \pm 5$$

respectively.

f does not change at all in any direction perpendicular to $3\mathbf{i} - 4\mathbf{k}$, i.e. any direction in the plane through $(-1, 0, 1)$ which is perpendicular to $3\mathbf{i} - 4\mathbf{k}$. If (x, y, z) is any point on this plane then its equation is

$$3(x + 1) - 4(z - 1) = 0$$

i.e.

$$3x - 4z + 7 = 0$$

So f does not increase at $(-1, 0, 1)$ in any direction in this plane.

1. In what directions does the function $f(x, y) = x^2 + 3y^2$ change most rapidly at $(0, 1)$? What are the directions of zero change in f at $(0, 1)$?

2. Find the directions in which the following functions increase and decrease most rapidly at the points indicated, and find the corresponding derivatives at these points:

   (i) $f(x, y, z) = x/y - yz$ at $(2, 1, 1)$

   (ii) $f(x, y, z) = x^2y - yz^3 + z$ at $(1, -2, 0)$

3. Discuss the relationship between the total differential and the directional derivative.

9.2 Tangent planes and normal lines

We introduced the tangent plane in Section 5.5. Here we look at it in the context of the gradient operator. Think of a mortar-board (the one you may be wearing one day!). The flat piece may be thought of as the *tangent plane* at the point to which it is attached on the curved piece. It clearly plays the same role for the surface as the tangent to a curve plays for a curve – it is the plane which, locally, touches the surface just once. But how do we formalize this in pure mathematical terms? There are a number of ways in which this might be done, but the most elementary is to characterize the tangent plane as composed of all the tangent lines to all the curves lying in the surface and passing through the tangent point. Here again we are essentially dealing with higher-dimensional situations by reducing them to lower-dimensional ones.

So, consider a level surface in three dimensions defined by $f(x, y, z) = c$, and let $\mathbf{r}(t) = x(t)\mathbf{i} + y(t)\mathbf{j} + z(t)\mathbf{k}$ be a smooth curve on the surface. Let $f(x, y, z)$ be differentiable. On the curve we have $f(x(t), y(t), z(t)) = c$. Differentiating both sides of this equation with respect to t gives

$$\frac{d}{dt} f(x(t), y(t), z(t))$$

$$= \frac{\partial f}{\partial x}\frac{dx}{dt} + \frac{\partial f}{\partial y}\frac{dy}{dt} + \frac{\partial f}{\partial z}\frac{dz}{dt}$$

$$\equiv \left(\frac{\partial f}{\partial x}\mathbf{i} + \frac{\partial f}{\partial y}\mathbf{j} + \frac{\partial f}{\partial z}\mathbf{k}\right) \cdot \left(\frac{dx}{dt}\mathbf{i} + \frac{dy}{dt}\mathbf{j} + \frac{dz}{dt}\mathbf{k}\right)$$

$$\equiv \nabla f \cdot \frac{d\mathbf{r}}{dt} = 0$$

Again remember that we still have to confirm that ∇f is a vector.

This relation between the gradient and the derivative of curves in a surface holds at all points. Now let us focus on one point P and consider all curves passing through this point. Then all the tangent vectors at P are orthogonal to ∇f at P, so the curves' tangent lines all lie in the plane through P normal to ∇f. *This* is the 'tangent plane' to the surface P. The line through P, perpendicular to the plane, is the *normal* to the surface at P.

This discussion motivates the following definitions. The *tangent plane* at the point $P_0(x_0, y_0, z_0)$ on the level surface $f(x, y, z) = c$ is the plane through P_0 normal to $\nabla f|_{P_0}$. The *normal line* of the surface at P_0 is the line through P_0 parallel to $\nabla f|_{P_0}$.

It follows that in terms of coordinates, the tangent plane has the equation

$$f_x(P_0)(x - x_0) + f_y(P_0)(y - y_0) + f_z(P_0)(z - z_0) = 0$$

as anticipated in Section 5.5. The normal line is described by the equations

$$x = x_0 + f_x(P_0)t \qquad y = y_0 + f_y(P_0)t \qquad z = z_0 + f_z(P_0)t$$

PROBLEM 5

Find the tangent plane and normal line of the surface

$$f(x, y, z) = x^2 - y^2 + z^2 + 2xy - 3xz = 0$$

at the point $P_0(1, 1, 1)$.

The gradient is

$$\nabla f = (2x + 2y - 3z)\mathbf{i} + (-2y + 2x)\mathbf{j} + (2z - 3x)\mathbf{k}$$

So at $(1, 1, 1)$

$$\nabla f|_{P_0} = \mathbf{i} - \mathbf{k}$$

and the tangent plane is

$$1.(x - 1) + (-1).(z - 1) = 0$$

or

$$x - z = 0$$

The normal to the surface at $(1, 1, 1)$ is

$$x = 1 + t \qquad y = 1 \qquad z = 1 - t$$

We often want to find tangent planes to the surface $z = f(x, y)$. These can be found simply by writing it in the form

$$F(x, y, z) = f(x, y) - z = 0$$

PROBLEM 6

Obtain the equations of the tangent plane to the surface $z = f(x, y)$ at the point $P_0(x_0, y_0, z_0)$. Find the plane tangent to the surface $z = e^x \cos y - y^2$ at $(0, 0, 1)$.

The gradient of $F = f(x, y) - z$ is

$$\nabla F = f_x(x, y)\mathbf{i} + f_y(x, y)\mathbf{j} - \mathbf{k}$$

So the equation of the tangent plane at $P_0(x_0, y_0, z_0)$ is

$$f_x(P_0)(x - x_0) + f_y(P_0)(y - y_0) - (z - z_0) = 0$$

For the surface $z = e^x \cos y - y^2$ we have $f_x = e^x \cos y, f_y = -e^x \sin y - 2y$, so the equation of the plane at $(0, 0, 1)$ is

$$1.(x - 0) + 0.(y - 0) - (z - 1) = 0$$

or

$$x - z + 1 = 0$$

Remember from Section 5.5 that we can write the tangent plane to the surface $z = f(x, y)$ at $P_0(x_0, y_0, z_0)$ in the form

$$z = z_0 + f_x(P_0)(x - x_0) + f_y(P_0)(y - y_0)$$

and so we may view the *linearization of $f(x, y)$* as a *tangent plane approximation* at P_0. This is exactly analogous to the way in which a tangent at a point on a curve is a local linear approximation to the curve at that point.

EXERCISES ON 9.2

1. Find the tangent planes and normal lines to the following surfaces at the points indicated:

    (i) $x^2 + y^2 + z^2 = 1$ at $\left(\dfrac{1}{4}, \dfrac{1}{4}, \dfrac{1}{\sqrt{2}}\right)$

    (ii) $\cos x \cos y \cos z = 1$ at $(0, \pi/2, \pi/3)$

2. Find the tangent planes to the following surfaces at the points indicated:

    (i) $z = x^2 + y^2$ $(0, 1, 1)$

    (ii) $z = e^{x+y}$ $(0, 0, 1)$

9.3 Differentiability revisited via the gradient

We cannot pass on without revisiting the idea of differentiability and expressing this in vector terms. The gradient then pops up quite naturally. All we really need to do is express the definition of differentiability given in Chapter 5 in corresponding vector form. I will simply give the definition and leave it as an exercise for you to make the connection. Let us go straight to $\mathbb{R}^3$.

Let f be a function of three variables that is defined in a neighbourhood of $\mathbf{r} = (x, y, z)$, and let $\Delta\mathbf{r} = (\Delta x, \Delta y, \Delta z)$. If $f_x(x, y, z), f_y(x, y, z), f_z(x, y, z)$ exist, then f is *differentiable* at $\mathbf{r}$ if there is a function g such that

$$f(\mathbf{r} + \Delta\mathbf{r}) - f(\mathbf{r}) = \nabla f . \Delta\mathbf{r} + g(\Delta\mathbf{r})$$

where

$$\lim_{|\Delta\mathbf{r}| \to 0} \frac{g(\Delta\mathbf{r})}{|\Delta\mathbf{r}|} = 0$$

Note how the gradient comes in here. Its occurrence is simply related to the fact that the tangent plane at a point provides a linear approximation to the surface at a point, and it is the existence of this approximation which guarantees the differentiability.

Show how the definition of differentiability given in this section is equivalent to that given in Section 5.2.

9.4 What is a vector – again?

You may recall that in elementary calculus it is often useful to regard the derivative of a function, dy/dx, as the effect of an *operator* acting on y, the D-operator:

$$Dy \equiv \left(\frac{d}{dx}\right) y = \frac{dy}{dx}$$

The advantage of this is that it allows differentiation and integration to be viewed as a sort of 'algebra' of the D-operator. It is a very useful device in solving differential equations, for example (Cox, 1996).

In D-operator terminology, we become used to manipulating with the object d/dx in a similar way to how we would with any other algebraic symbol. So, we might expect to be able to do a similar thing with partial derivatives, such as $\partial/\partial x$, $\partial/\partial y$, $\partial/\partial z$. In fact, we used an example of this as a useful trick in Section 4.5. Here we want to extend it. In particular, perhaps we could form a 'vector' with these objects as components – let us denote it by

$$\nabla \equiv \left(\frac{\partial}{\partial x}, \frac{\partial}{\partial y}, \frac{\partial}{\partial z}\right)$$

In fact, we can – but we have to check that it *is* a vector, i.e. that it transforms in the manner required of a vector under the translation and rotation of axes (*see* Section 8.4). But let us suppose this has been done. Then we still have to remember that ∇ is a derivative operator, which can act on functions of x, y, z. These functions may themselves be vectors. We can get some idea of the sorts of things we can get by treating ∇ like an ordinary vector, but one which does not necessarily *commute* with other vectors. Thus, if $\phi(x, y, z)$ is an arbitrary differentiable function of (x, y, z) then we could form

$$\phi\nabla \equiv \left(\phi\frac{\partial}{\partial x}, \phi\frac{\partial}{\partial y}, \phi\frac{\partial}{\partial z}\right)$$

or

$$\nabla\phi \equiv \left(\frac{\partial\phi}{\partial x}, \frac{\partial\phi}{\partial y}, \frac{\partial\phi}{\partial z}\right)$$

which are quite different objects. Also, if $\mathbf{F}(x, y, z)$ is a vector function of x, y, z then we could form objects such as $\mathbf{F} \cdot \nabla$, $\nabla \cdot \mathbf{F}$, $\mathbf{F} \times \nabla$, $\nabla \times \mathbf{F}$, or $\nabla \times (\nabla \times \mathbf{F})$. The only question which remains is whether such objects are useful – and of course they are! They and their properties form the subject of this chapter.

Show that
(i) $\nabla(\phi + \psi) = \nabla\phi + \nabla\psi$
(ii) $\nabla(\phi\psi) = (\nabla\phi)\psi + \phi(\nabla\psi)$

9.5 Scalar and vector fields

We are going to be dealing with functions of variables (x, y, z) which have the additional properties that they behave as scalars or vectors under translation and rotation of rectangular coordinate axes. In general, a function $\phi(x, y, z)$, for fixed values of (x, y, z), may be a *constant* – but that does not necessarily mean that it is a *scalar*. It may, for example, be the third component of a vector – and under coordinate transformations it may not transform as a *scalar*. Similarly, a three-component object, $(F_1(x, y, z)\,,\ F_2(x, y, z),\ F_3(x, y, z))$, is not necessarily a vector.

However, since we are seeking to develop a vector calculus, we want all the functions that we deal with to be of a scalar or vector nature. This leads us to the definition of scalar and vector fields.

● *Scalar and vector fields* ───────────────────────────

Let $\phi(x, y, z)$ be a scalar defined on an appropriate domain in three-dimensional space (note – this does not have to be physical space, of course!). In other words, to each point $P(x, y, z)$ in the domain we assign a scalar quantity $\phi(x, y, z)$. Then we say that ϕ is a *scalar field*. We can write it succinctly as $\phi(\mathbf{r})$, where $\mathbf{r}$ denotes the position vector referred to the origin of coordinates.

Let $\mathbf{F}(x, y, z)$ be a vector defined on a domain of 3-space. Then we say $\mathbf{F}$ is a *vector field*, or *vector function of position*. Again, we can write it as $\mathbf{F}(\mathbf{r})$.

In most cases the domain we refer to will in fact be a region of 3-space.

PROBLEM 7

> The force on a body which is at a distance r from the centre of the Earth has magnitude inversely proportional to the square of the distance r and is directed towards the Earth's surface. Specify the vector field and the region of space over which it is defined.

In this case the vector field is the Earth's gravitational force acting at all points in a region outside of the Earth's radius. If we take rectangular coordinates (x, y, z) with origin 0, at the centre of the Earth, and model the Earth as a sphere of radius R, then this *force field* acts at all points (x, y, z) in space such that

$$\sqrt{x^2 + y^2 + z^2} \geq R$$

Let $\mathbf{e}_r$ be a unit vector in the direction $\overrightarrow{0P}$. Then the vector field function acting at point P is $-F_g\mathbf{e}_r$, where the magnitude F_g is proportional to $1/(0P)^2$, so

$$F_g = \frac{K}{0P^2} = \frac{K}{x^2 + y^2 + z^2}$$

for some positive constant K. The direction of the force field vector $\mathbf{F}$ is along the vector

$$\overrightarrow{P0} = -(x\mathbf{i} + y\mathbf{j} + z\mathbf{k})$$

and so we can write

$$\mathbf{F} = -F_g \frac{\vec{OP}}{|\vec{OP}|} = \frac{-K}{x^2 + y^2 + z^2} \frac{(x\mathbf{i} + y\mathbf{j} + z\mathbf{k})}{\sqrt{x^2 + y^2 + z^2}}$$

$$= \frac{-K(x\mathbf{i} + y\mathbf{j} + z\mathbf{k})}{(x^2 + y^2 + z^2)^{3/2}} \qquad \text{for } \sqrt{x^2 + y^2 + z^2} \geq R$$

$$= \frac{-K\mathbf{r}}{r^{3/2}} \qquad r \geq R$$

This is a three-dimensional vector field defined at points in a three-dimensional space. Although such physical examples are useful in motivating vector fields, all we are really concerned about in this book is the fact that a vector field represents a vector defined at each point of a region of two- or (mostly) three-dimensional space. If we are using Cartesian rectangular coordinates then we can always write (in the three-dimensional case) such a vector field in the form

$$\mathbf{F}(x, y, z) = F_1(x, y, z)\mathbf{i} + F_2(x, y, z)\mathbf{j} + F_3(x, y, z)\mathbf{k}$$

EXERCISE ON 9.5

We can draw sketches of vector fields by placing at typical points arrows which have lengths proportional to the magnitude of the vector field at those points and oriented in the direction of the vector field at the points. Sketch the two-dimensional fields

(i) $\mathbf{F} = K(\mathbf{i} + \mathbf{j})$, where K is a constant

(ii) $\mathbf{F} = \dfrac{x\mathbf{i} + y\mathbf{j}}{\sqrt{x^2 + y^2}}$

Also sketch the three-dimensional gravitational field discussed in this section.

9.6 The gradient of a scalar field

Let $\phi(x, y, z)$ be a continuously differentiable scalar field defined on a region R, with x, y, z rectangular coordinates. Then the vector operator

$$\text{grad } \phi \equiv \nabla\phi = \frac{\partial\phi}{\partial x}\mathbf{i} + \frac{\partial\phi}{\partial y}\mathbf{j} + \frac{\partial\phi}{\partial z}\mathbf{k} \equiv \left(\frac{\partial\phi}{\partial x}, \frac{\partial\phi}{\partial y}, \frac{\partial\phi}{\partial z}\right)$$

is a vector field on R, and is called the *gradient* of ϕ. The operator

$$\nabla = \left(\frac{\partial}{\partial x}, \frac{\partial}{\partial y}, \frac{\partial}{\partial z}\right)$$

is called the *del operator*; we introduced this in Section 9.4.

PROBLEM 8

Find the gradient of the following scalar fields:

(i) $\phi = r = \sqrt{x^2 + y^2 + z^2}$

(ii) $\phi = [1 - (x^2 + y^2 + z^2)]^{1/2}$

(i) We have

$$\frac{\partial \phi}{\partial x} = \frac{x}{r} \quad \frac{\partial \phi}{\partial y} = \frac{y}{r} \quad \frac{\partial \phi}{\partial z} = \frac{z}{r}$$

So

$$\nabla \phi = \left(\frac{x}{r}, \frac{y}{r}, \frac{z}{r} \right)$$

(ii) $$\frac{\partial \phi}{\partial x} = \frac{-x}{[1 - (x^2 + y^2 + z^2)]^{1/2}} \qquad \frac{\partial \phi}{\partial y} = \frac{-y}{[1 - (x^2 + y^2 + z^2)]^{1/2}}$$

$$\frac{\partial \phi}{\partial z} = \frac{-z}{[1 - (x^2 + y^2 + z^2)]^{1/2}}$$

So

$$\nabla \phi = \frac{1}{[1 - (x^2 + y^2 + z^2)]^{1/2}} (-x, -y, -z)$$

We will now confirm that $\nabla \phi$ is in fact a vector field on the region R by verifying that the components are invariant under a translation of axes and transform as a vector under a rotation of axes.

PROBLEM 9

Verify invariance under translation.

Using a translation of axes, the coordinates transform according to

$$x = x' + a \quad y = y' + b \quad z = z' + c$$

say. Then the chain rule gives

$$\frac{\partial \phi}{\partial x'} = \frac{\partial x}{\partial x'} \frac{\partial \phi}{\partial x} + \frac{\partial y}{\partial x'} \frac{\partial \phi}{\partial y} + \frac{\partial z}{\partial x'} \frac{\partial \phi}{\partial z}$$

$$= \frac{\partial \phi}{\partial x}$$

and similarly

$$\frac{\partial \phi}{\partial y'} = \frac{\partial \phi}{\partial y} \qquad \frac{\partial \phi}{\partial z'} = \frac{\partial \phi}{\partial z}$$

Thus, the components of $\nabla \phi$ are invariant under a translation of coordinate axes.

Now consider a rotation of axes, under which the coordinates transform according to (note that we need to use the *inverse* transformation of the coordinates to get the *direct* transformation of the derivatives)

$$x = \ell_{11} x' + \ell_{21} y' + \ell_{31} z'$$
$$y = \ell_{12} x' + \ell_{22} y' + \ell_{32} z'$$
$$z = \ell_{13} x' + \ell_{23} y' + \ell_{33} z'$$

PROBLEM 10

> Determine how the components of $\nabla\phi$ transform under such a transformation.

We use the chain rule again. For example,

$$\frac{\partial\phi}{\partial x'} = \frac{\partial x}{\partial x'}\frac{\partial\phi}{\partial x} + \frac{\partial y}{\partial x'}\frac{\partial\phi}{\partial y} + \frac{\partial z}{\partial x'}\frac{\partial\phi}{\partial z}$$

$$= \ell_{11}\frac{\partial\phi}{\partial x} + \ell_{12}\frac{\partial\phi}{\partial y} + \ell_{13}\frac{\partial\phi}{\partial z}$$

and similarly

$$\frac{\partial\phi}{\partial y'} = \ell_{21}\frac{\partial\phi}{\partial x} + \ell_{22}\frac{\partial\phi}{\partial y} + \ell_{23}\frac{\partial\phi}{\partial z}$$

$$\frac{\partial\phi}{\partial z'} = \ell_{31}\frac{\partial\phi}{\partial x} + \ell_{32}\frac{\partial\phi}{\partial y} + \ell_{33}\frac{\partial\phi}{\partial z}$$

This demonstrates that the three components of $\nabla\phi$ transform like a vector under rotations. So $\nabla\phi$ is indeed a vector field. Note that because of the use of the chain rule we need more than just the existence of the derivatives of ϕ. It is sufficient, however, to demand continuous differentiability of ϕ.

We have already discussed the geometric properties of the gradient of a scalar field in Sections 9.1 and 9.2. Here we will focus on its operational properties – and in particular we will use the del operator ∇ to define new vector differential operators.

First notice that since ∇ is a differential operator it follows immediately that for any two scalar fields, ϕ, ψ,

$$\nabla(\phi + \psi) = \nabla\phi + \nabla\psi$$

i.e. ∇ is a *linear operator*. In addition,

$$\nabla(\phi\psi) = \phi\nabla\psi + \psi\nabla\phi$$

i.e. the product rule applies.

It sometimes happens that we need to reconstruct a scalar field ϕ given its gradient – i.e. we have to solve an equation of the form $\nabla\phi = \mathbf{v}$, with $\mathbf{v}$ a given vector field. This is equivalent to integrating the three equations

$$\frac{\partial\phi}{\partial x} = v_1(x, y, z) \qquad \frac{\partial\phi}{\partial y} = v_2(x, y, z) \qquad \frac{\partial\phi}{\partial z} = v_3(x, y, z)$$

Integration of any one of these equations will involve an arbitrary function of the remaining two variables, giving three arbitrary functions. It simply remains to choose these functions to give a consistent solution for ϕ which satisfies all three equations (*see* Exercise 9.6.3). An obvious question is whether such a ϕ exists for any vector field $\mathbf{F}$. The answer is no: $\mathbf{F}$ must be a particular kind of vector field (*see* Exercise 9.7.4). We will revisit this issue in Section 10.6.

EXERCISES ON 9.6

1. Find the gradients of the following two- and three-dimensional scalar fields:
   (i) $\phi(x, y) = x^2 + 3y^2$
   (ii) $\phi(x, y, z) = (x + y + z)^2$
   (iii) $\phi(x, y) = \cos(xy)$
   (iv) $\phi(x, y, z) = ze^x \cos y$
   (v) $\phi(x, y, z) = \dfrac{x - z}{\sqrt{1 - y^2 + x^2}}$

2. Obtain the gradient of the scalar field

   $$F(x, y) = \ln\left[\frac{\sqrt{(x-1)^2 + y^2}}{\sqrt{(x+1)^2 + y^2}}\right]$$

   This force field is in fact due to the electric field in the plane caused by two infinite straight uniformly charged wires of equal and opposite charge that are perpendicular to the xy-plane and pass through the points $(1, 0)$ and $(-1, 0)$.

3. Find a scalar field ϕ such that $\nabla\phi = yz\mathbf{i} + xz\mathbf{j} + xy\mathbf{k}$.

9.7 Divergence and curl of a vector field

If we just think of the vector character of the gradient – i.e. regard ∇ as a vector – then we are led to think about the sorts of things we might combine with it. We can multiply it on either side by a scalar, as in Section 9.4. We can also take advantage of the scalar product and the vector product to multiply it by a vector. Thus, let $\mathbf{F}(x, y, z) = (F_1, F_2, F_3)$ be a vector field. We could then form $\mathbf{F} \cdot \nabla$, $\nabla \cdot \mathbf{F}$, $\mathbf{F} \times \nabla$, or $\nabla \times \mathbf{F}$. From a purely vectorial point of view these have the component forms

$$\mathbf{F} \cdot \nabla \equiv F_1\frac{\partial}{\partial x} + F_2\frac{\partial}{\partial y} + F_3\frac{\partial}{\partial z}$$

$$\nabla \cdot \mathbf{F} \equiv \frac{\partial F_1}{\partial x} + \frac{\partial F_2}{\partial y} + \frac{\partial F_3}{\partial z}$$

$$\mathbf{F} \times \nabla \equiv \mathbf{i}\left(F_3\frac{\partial}{\partial y} - F_2\frac{\partial}{\partial z}\right) + \mathbf{j}\left(F_1\frac{\partial}{\partial z} - F_3\frac{\partial}{\partial x}\right) + \mathbf{k}\left(F_2\frac{\partial}{\partial x} - F_1\frac{\partial}{\partial y}\right)$$

$$\nabla \times \mathbf{F} \equiv \mathbf{i}\left(\frac{\partial F_3}{\partial y} - \frac{\partial F_2}{\partial z}\right) + \mathbf{j}\left(\frac{\partial F_1}{\partial z} - \frac{\partial F_3}{\partial x}\right) + \mathbf{k}\left(\frac{\partial F_2}{\partial x} - \frac{\partial F_1}{\partial y}\right)$$

$\mathbf{F} \cdot \nabla$ and $\mathbf{F} \times \nabla$ are still vector differential operators. $\nabla \cdot \mathbf{F}$, on the other hand, is a scalar function or *scalar field*, called the *divergence* of the vector field $\mathbf{F}$. $\nabla \times \mathbf{F}$ is a *vector field* called the *curl* of the vector field $\mathbf{F}$. Both the divergence and curl are important objects in vector calculus. We write

$$\nabla \cdot \mathbf{F} \equiv \text{div } \mathbf{F} = \frac{\partial F_1}{\partial x} + \frac{\partial F_2}{\partial y} + \frac{\partial F_3}{\partial z}$$

$$\nabla \times \mathbf{F} \equiv \text{curl } \mathbf{F} = \left(\frac{\partial F_3}{\partial y} - \frac{\partial F_2}{\partial z}\right)\mathbf{i} + \left(\frac{\partial F_1}{\partial z} - \frac{\partial F_3}{\partial x}\right)\mathbf{j} + \left(\frac{\partial F_2}{\partial x} - \frac{\partial F_1}{\partial y}\right)\mathbf{k}$$

If you have trouble remembering the form of the curl, you may prefer the 'determinantal' expression

$$\text{curl } \mathbf{F} = \begin{vmatrix} \mathbf{i} & \mathbf{j} & \mathbf{k} \\ \dfrac{\partial}{\partial x} & \dfrac{\partial}{\partial y} & \dfrac{\partial}{\partial z} \\ F_1 & F_2 & F_3 \end{vmatrix}$$

We said that div $\mathbf{F}$ is a scalar field. This may be easily seen by noting that ∇ is itself a vector (the ϕ in the proof in Section 9.6 is irrelevant in proving that $\nabla\phi$ is a vector, as you can easily check) and so $\nabla \cdot \mathbf{F}$ is just the scalar product of two vectors. However, it is instructive to prove the result formally. Invariance of $\nabla \cdot \mathbf{F}$ under translations is clear from the fact that both ∇ and $\mathbf{F}$ are separately invariant. For rotations, it helps to introduce subscript notation (this is effectively a prelude to tensor notation). Thus, denote x, y, z, by x_1, x_2, x_3, so that, using the summation convention,

$$\nabla \cdot \mathbf{F} \equiv \frac{\partial F_1}{\partial x_1} + \frac{\partial F_2}{\partial x_2} + \frac{\partial F_3}{\partial x_3} \equiv \frac{\partial F_i}{\partial x_i}$$

Now consider a rotation to new axes $0x_1'x_2'x_3'$. The coordinates transform according to

$$x_i' = \ell_{ij} x_j$$
$$x_i = \ell_{ji} x_j'$$

with the direction cosines ℓ_{ij} satisfying the orthonormality conditions

$$\ell_{ik} \ell_{jk} = \delta_{ij}$$

Then the partial derivatives transform according to

$$\frac{\partial}{\partial x_j'} = \ell_{ji} \frac{\partial}{\partial x_i}$$

while the vector components F_i transform according to

$$F_j' = \ell_{jk} F_k$$

Since the ℓ_{ij} are independent of the x_i, we then have

$$\frac{\partial F_j'}{\partial x_j'} = \ell_{ji}\ell_{jk} \frac{\partial F_k}{\partial x_i}$$

$$= \delta_{ik} \frac{\partial F_k}{\partial x_i} = \frac{\partial F_i}{\partial x_i}$$

In other words,

$$\frac{\partial F_1'}{\partial x_1'} + \frac{\partial F_2'}{\partial x_2'} + \frac{\partial F_3'}{\partial x_3'} = \frac{\partial F_1}{\partial x_1} + \frac{\partial F_2}{\partial x_2} + \frac{\partial F_3}{\partial x_3}$$

Thus, div $\mathbf{F}$ is invariant under a rotation of axes, and we have completed the proof that it is a scalar field.

Again, we can see directly that $\nabla \times \mathbf{F}$, being a vector product of two vectors, is itself also a vector. This may also be proved similarly to the case of div $\mathbf{F}$ above – we leave this as an exercise.

PROBLEM 11

> Find the divergence and curl of each of the following vector fields:
> (i) $\mathbf{F} = xy\mathbf{i} + yz\mathbf{j} + xz\mathbf{k}$ (ii) $\mathbf{G} = x\mathbf{i} + y\mathbf{j} + z\mathbf{k}$

(i) We have

$$\operatorname{div} \mathbf{F} = \nabla . \mathbf{F} = \frac{\partial}{\partial x}(xy) + \frac{\partial}{\partial y}(yz) + \frac{\partial}{\partial z}(xz)$$

$$= y + z + x$$

$$\operatorname{curl} \mathbf{F} = \nabla \times \mathbf{F} = \left(\frac{\partial(xz)}{\partial y} - \frac{\partial(yz)}{\partial z}\right)\mathbf{i} + \left(\frac{\partial(xy)}{\partial z} - \frac{\partial(xz)}{\partial x}\right)\mathbf{j} + \left(\frac{\partial(yz)}{\partial x} - \frac{\partial(xy)}{\partial y}\right)\mathbf{k}$$

$$= -y\mathbf{i} - z\mathbf{j} - x\mathbf{k}$$

(ii) $\mathbf{G} = x\mathbf{i} + y\mathbf{j} + z\mathbf{k}$
So

$$\operatorname{div} \mathbf{G} = \frac{\partial(x)}{\partial x} + \frac{\partial(y)}{\partial y} + \frac{\partial(z)}{\partial z} = 3$$

$$\operatorname{curl} \mathbf{G} = \left(\frac{\partial(z)}{\partial y} - \frac{\partial(y)}{\partial z}\right)\mathbf{i} + \left(\frac{\partial(x)}{\partial z} - \frac{\partial(z)}{\partial x}\right)\mathbf{j} + \left(\frac{\partial(y)}{\partial x} - \frac{\partial(x)}{\partial y}\right)\mathbf{k}$$

$$\equiv 0$$

EXERCISES ON 9.7

1. Prove that $\nabla \times \mathbf{F}$ transforms as a vector under rotations of the axes.
2. Evaluate (a) div (b) curl (c) grad(div) for the following vector fields:
    (i) $\mathbf{F} = x\mathbf{i} + 2y\mathbf{j} + z\mathbf{k}$ (ii) $\mathbf{F} = e^x\mathbf{i} + e^y\mathbf{j} + e^z\mathbf{k}$
    (iii) $\mathbf{F} = F_1(x)\mathbf{i} + F_2(y)\mathbf{j} + F_3(z)\mathbf{k}$ (iv) $\mathbf{F} = \cos(xy)\mathbf{i} + \sin(yz)\mathbf{j}$
    Why does this question not include any other combinations of div, grad and curl?
3. Prove that the vector field $\mathbf{H} = \phi\nabla\psi$ is perpendicular to curl $\mathbf{H}$ at all points where neither vector field vanishes.
4. Show that if $\mathbf{v} = \nabla\phi$ then curl $\mathbf{v} = 0$.

9.8 Properties of div and curl: vector identities

This section is really a game. Now that we have defined all these new objects, grad, div and curl, we can enjoy ourselves applying them to scalar and vector fields, to each other, and generally studying their properties. There is little you cannot deduce for yourself with a bit of help, so this section consists mainly of worked problems and exercises – and I shall have very little to say.

It is easy to see that div and curl, like grad, are linear operators:

$$\operatorname{div}(\mathbf{F} + \mathbf{G}) = \operatorname{div} \mathbf{F} + \operatorname{div} \mathbf{G}$$
$$\operatorname{curl}(\mathbf{F} + \mathbf{G}) = \operatorname{curl} \mathbf{F} + \operatorname{curl} \mathbf{G}$$

for arbitrary vector fields $\mathbf{F}$ and $\mathbf{G}$.

Given a vector field **F** we can evaluate things like grad(div **F**), or for a scalar field ϕ things like div(grad ϕ), to our hearts' content. We only have to check consistency of the vector combinations – for example, the curl of a scalar field is not defined.

PROBLEM 12

Evaluate
(i) grad(div **F**)
(ii) div(curl **F**)
for the vector field $\mathbf{F} = yz\mathbf{j} + zx\mathbf{k}$.

$$\text{div } \mathbf{F} = \frac{\partial(yz)}{\partial y} + \frac{\partial(zx)}{\partial z} = x + z$$

$$\text{curl } \mathbf{F} = \begin{vmatrix} \mathbf{i} & \mathbf{j} & \mathbf{k} \\ \frac{\partial}{\partial x} & \frac{\partial}{\partial y} & \frac{\partial}{\partial z} \\ 0 & yz & zx \end{vmatrix} = -y\mathbf{i} + z\mathbf{j}$$

So

(i) $\text{grad(div } \mathbf{F}) = \frac{\partial(x+z)}{\partial x}\mathbf{i} + \frac{\partial(x+z)}{\partial y}\mathbf{j} + \frac{\partial(x+z)}{\partial z}\mathbf{k}$

$$= \mathbf{i} + \mathbf{k}$$

(ii) $\text{div(curl } \mathbf{F}) = \frac{\partial(-y)}{\partial x} + \frac{\partial(z)}{\partial y} = 0$

From these two routine examples a couple of questions arise. First, what do the component forms of these second-order derivative operators look like – and can they be expressed in alternative, possibly simpler forms? Second, is the zero result in (ii) a coincidence, or is it always true that div(curl **F**) = 0?

The answer to the second question is yes, and it actually provides an example of the sort of thing which provides an answer to the first question. We can in fact find many identities between the new objects we have defined. Before we delve into these, let us look at one very important operator which arises just about everywhere in the theory of fields, partial differential equations and so on.

PROBLEM 13

Write $\nabla^2\phi = \nabla.(\nabla\phi) = \text{div(grad } \phi)$ in component form.

First, note that both the expression and the notation make sense. $\nabla\phi$ is a vector, and so we can take its divergence $\nabla.(\nabla\phi)$. And if we were talking about ordinary vectors then $\mathbf{a} \cdot \mathbf{a} = a^2$ is the notation we would use – it denotes the square of the magnitude of $\mathbf{a}$. So ∇^2 makes sense – we might think of it as the square of the magnitude of ∇.

Now, we have

$$\nabla^2\phi = \nabla\cdot(\nabla\phi)$$

$$= \nabla\cdot\left(\frac{\partial\phi}{\partial x}\mathbf{i} + \frac{\partial\phi}{\partial y}\mathbf{j} + \frac{\partial\phi}{\partial z}\mathbf{k}\right)$$

$$= \frac{\partial}{\partial x}\left(\frac{\partial\phi}{\partial x}\right) + \frac{\partial}{\partial y}\left(\frac{\partial\phi}{\partial y}\right) + \frac{\partial}{\partial z}\left(\frac{\partial\phi}{\partial z}\right)$$

$$= \frac{\partial^2\phi}{\partial x^2} + \frac{\partial^2\phi}{\partial y^2} + \frac{\partial^2\phi}{\partial z^2}$$

Lifting the ∇^2 free of the scalar field ϕ, we call the second-order derivative operator

$$\nabla^2 \equiv \frac{\partial^2}{\partial x^2} + \frac{\partial^2}{\partial y^2} + \frac{\partial^2}{\partial z^2}$$

the *Laplacian,* after the French mathematician Pierre Simon de Laplace (1749–1827). Using the previously introduced subscript notation we can write ∇^2 as

$$\nabla^2 \equiv \frac{\partial^2}{\partial x_1^2} + \frac{\partial^2}{\partial x_2^2} + \frac{\partial^2}{\partial x_3^2} \equiv \frac{\partial}{\partial x_i}\frac{\partial}{\partial x_i}$$

using the summation convention.

Again, it is instructive to prove invariance of ∇^2 directly, even though it follows from the fact that ∇ is a vector. Invariance of ∇^2 under translations follows from the invariance of $\partial/\partial x_i$. Under rotations we have

$$\frac{\partial}{\partial x_j'} = \ell_{ji}\frac{\partial}{\partial x_i}$$

so

$$\frac{\partial}{\partial x_j'}\frac{\partial}{\partial x_j'} = \left(\ell_{ji}\frac{\partial}{\partial x_i}\right)\left(\ell_{jk}\frac{\partial}{\partial x_k}\right)$$

$$= \delta_{ik}\frac{\partial}{\partial x_i}\frac{\partial}{\partial x_k}$$

$$= \frac{\partial}{\partial x_i}\frac{\partial}{\partial x_i}$$

demonstrating the invariance of ∇^2 under rotations and completing the proof that it is a scalar operator.

Since ∇^2 is a scalar, then it can operate on either a scalar or a vector. Operation on a scalar field is straightforward: we know that

$$\nabla^2\phi \equiv \mathrm{div}(\mathrm{grad}\ \phi)$$

On a vector field $\mathbf{F}$, grad $\mathbf{F}$ is not defined, so it is not immediately clear how ∇^2 acts on $\mathbf{F}$ in vector terms. To gain insight into this, attempt the following problem.

PROBLEM 14

> Obtain the component forms of the following expressions: (i) grad(div **F**),
> (ii) curl curl **F**, where **F** is a continuously differentiable vector field.

(i) We have, with the usual notation

$$\text{grad(div } \mathbf{F}) = \nabla\left(\frac{\partial F_1}{\partial x_1} + \frac{\partial F_2}{\partial x_2} + \frac{\partial F_3}{\partial x_3}\right)$$

$$= \left(\frac{\partial^2 F_1}{\partial x_1^2} + \frac{\partial^2 F_2}{\partial x_1 \partial x_2} + \frac{\partial^2 F_3}{\partial x_1 \partial x_3}\right)\mathbf{i}$$

$$+ \left(\frac{\partial^2 F_1}{\partial x_2 \partial x_1} + \frac{\partial^2 F_2}{\partial x_2^2} + \frac{\partial^2 F_3}{\partial x_2 \partial x_3}\right)\mathbf{j}$$

$$+ \left(\frac{\partial^2 F_1}{\partial x_3 \partial x_1} + \frac{\partial^2 F_2}{\partial x_3 \partial x_2} + \frac{\partial^2 F_3}{\partial x_3^2}\right)\mathbf{k}$$

(ii)   curl curl $\mathbf{F} \equiv \begin{vmatrix} \mathbf{i} & \mathbf{j} & \mathbf{k} \\ \dfrac{\partial}{\partial x_1} & \dfrac{\partial}{\partial x_2} & \dfrac{\partial}{\partial x_3} \\ \dfrac{\partial F_3}{\partial x_2} - \dfrac{\partial F_2}{\partial x_3} & \dfrac{\partial F_1}{\partial x_3} - \dfrac{\partial F_3}{\partial x_1} & \dfrac{\partial F_2}{\partial x_1} - \dfrac{\partial F_1}{\partial x_2} \end{vmatrix}$

$$= \left(\frac{\partial^2 F_2}{\partial x_2 \partial x_1} - \frac{\partial^2 F_1}{\partial x_2^2} - \frac{\partial^2 F_1}{\partial x_3^2} + \frac{\partial^2 F_3}{\partial x_3 \partial x_1}\right)\mathbf{i}$$

$$- \left(\frac{\partial^2 F_2}{\partial x_1^2} - \frac{\partial^2 F_1}{\partial x_1 \partial x_2} - \frac{\partial^2 F_3}{\partial x_3 \partial x_2} + \frac{\partial^2 F_2}{\partial x_3^2}\right)\mathbf{j}$$

$$+ \left(\frac{\partial^2 F_1}{\partial x_1 \partial x_3} - \frac{\partial^2 F_3}{\partial x_1^2} - \frac{\partial^2 F_2}{\partial x_2^2} + \frac{\partial^2 F_2}{\partial x_2 \partial x_3}\right)\mathbf{k}$$

Subtracting the results of (ii) from (i), we see that all mixed derivatives cancel out (assuming reversal of order is permissible), and we obtain the identity

$$\nabla^2 \mathbf{F} \equiv \text{grad(div } \mathbf{F}) - \text{curl curl } \mathbf{F}$$

which shows us how ∇^2 may be applied to a vector field in terms of the usual vector differential operators.

The partial differential equation

$$\nabla^2 \phi = \frac{\partial^2 \phi}{\partial x^2} + \frac{\partial^2 \phi}{\partial y^2} + \frac{\partial^2 \phi}{\partial z^2} = 0$$

is called *Laplace's equation in three-dimensional rectangular coordinates*.

So far, we have given a number of identities involving vector differential operators. Below we list more of these – the most significant ones, which occur frequently. You are invited to prove most of them yourself – with some help. The operator notation equivalents are also given. In each case we assume all required conditions of continuity and differentiability are fulfilled.

(i) $\text{div}(\text{curl } \mathbf{F}) \equiv \nabla \cdot (\nabla \times \mathbf{F}) \equiv 0$

(ii) $\text{curl}(\text{grad } \phi) \equiv \nabla \times (\nabla \phi) \equiv 0$

(iii) $\text{grad}(\phi_1 \phi_2) \equiv \phi_1 \text{ grad } \phi_2 + \phi_2 \text{ grad } \phi_1$, or

$\quad\quad \nabla(\phi_1 \phi_2) \equiv \phi_1 \nabla \phi_2 + \phi_2 \nabla \phi_1$

(iv) $\text{div}(\phi \mathbf{F}) \equiv \phi \text{ div } \mathbf{F} + \mathbf{F} \cdot \text{grad } \phi$, or

$\quad\quad \nabla \cdot (\phi \mathbf{F}) \equiv \phi \nabla \cdot \mathbf{F} + \mathbf{F} \cdot \nabla \phi$

(v) $\text{curl}(\phi \mathbf{F}) \equiv \phi \text{ curl } \mathbf{F} - \mathbf{F} \times \text{grad } \phi$, or

$\quad\quad \nabla \times (\phi \mathbf{F}) \equiv \phi \nabla \times \mathbf{F} - \mathbf{F} \times \nabla \phi$

(vi) $\text{grad}(\mathbf{F} \cdot \mathbf{G}) \equiv \mathbf{F} \times \text{curl } \mathbf{G} + \mathbf{G} \times \text{curl } \mathbf{F} + (\mathbf{F} \cdot \nabla)\mathbf{G} + (\mathbf{G} \cdot \nabla)\mathbf{F}$, or

$\quad\quad \nabla(\mathbf{F} \cdot \mathbf{G}) \equiv \mathbf{F} \times (\nabla \times \mathbf{G}) + \mathbf{G} \times (\nabla \times \mathbf{F}) + (\mathbf{F} \cdot \nabla)\mathbf{G} + (\mathbf{G} \cdot \nabla)\mathbf{F}$

(vii) $\text{div}(\mathbf{F} \times \mathbf{G}) \equiv \mathbf{G} \cdot \text{curl } \mathbf{F} - \mathbf{F} \cdot \text{curl } \mathbf{G}$, or

$\quad\quad \nabla \cdot (\mathbf{F} \times \mathbf{G}) \equiv \mathbf{G} \cdot (\nabla \times \mathbf{F}) - \mathbf{F} \cdot (\nabla \times \mathbf{G})$

(viii) $\text{curl}(\mathbf{F} \times \mathbf{G}) \equiv \mathbf{F} \text{ div } \mathbf{G} - \mathbf{G} \text{ div } \mathbf{F} + (\mathbf{G} \cdot \nabla)\mathbf{F} - (\mathbf{F} \cdot \nabla)\mathbf{G}$, or

$\quad\quad \nabla \times (\mathbf{F} \times \mathbf{G}) \equiv \mathbf{F}(\nabla \cdot \mathbf{G}) - \mathbf{G}(\nabla \cdot \mathbf{F}) + (\mathbf{G} \cdot \nabla)\mathbf{F} - (\mathbf{F} \cdot \nabla)\mathbf{G}$

PROBLEM 15

> Prove (i) and (ii) using the component forms.

(i) This is straightforward:

$$\text{div}(\text{curl } \mathbf{F}) \equiv \frac{\partial}{\partial x}\left(\frac{\partial F_3}{\partial y} - \frac{\partial F_2}{\partial z}\right) + \frac{\partial}{\partial y}\left(\frac{\partial F_1}{\partial z} - \frac{\partial F_3}{\partial x}\right) + \frac{\partial}{\partial z}\left(\frac{\partial F_2}{\partial x} - \frac{\partial F_1}{\partial y}\right)$$

$$\equiv 0$$

Note that it is necessary to assume that the derivatives are continuous, so that we can reverse the order of differentiation.

(ii) $\text{curl}(\text{grad } \phi) \equiv \begin{vmatrix} \mathbf{i} & \mathbf{j} & \mathbf{k} \\ \dfrac{\partial}{\partial x} & \dfrac{\partial}{\partial y} & \dfrac{\partial}{\partial z} \\ \dfrac{\partial \phi}{\partial x} & \dfrac{\partial \phi}{\partial y} & \dfrac{\partial \phi}{\partial z} \end{vmatrix} \equiv 0$

again following trivially by reversing the order of derivatives.

The proof of the remaining identities can be done componentwise, as above, but it is much easier if we extend the subscript notation to the basis vectors, writing $\mathbf{e}_1 = \mathbf{i}$, $\mathbf{e}_2 = \mathbf{j}$, $\mathbf{e}_3 = \mathbf{k}$. Then the component expressions for grad, div and curl are very simple, since we can use the summation convention:

$$\text{grad } \phi = \mathbf{e}_i \frac{\partial \phi}{\partial x_i}$$

$$\text{div } \mathbf{F} = \mathbf{e}_i \cdot \frac{\partial \mathbf{F}}{\partial x_i}$$

$$\text{curl } \mathbf{F} = \mathbf{e}_i \times \frac{\partial \mathbf{F}}{\partial x_i}$$

in which

$$\frac{\partial \mathbf{F}}{\partial x_i} = \left(\frac{\partial F_1}{\partial x_i}, \frac{\partial F_2}{\partial x_i}, \frac{\partial F_3}{\partial x_i} \right)$$

The proof of these is left as an exercise.

PROBLEM 16

Prove (iii), (iv) using the above notation.

(iii) We have

$$\mathrm{grad}(\phi_1 \phi_2) \equiv \mathbf{e}_i \frac{\partial}{\partial x_i}(\phi_1 \phi_2)$$

$$\equiv \phi_1 \mathbf{e}_i \frac{\partial \phi_2}{\partial x_i} + \phi_2 \, \mathbf{e}_i \frac{\partial \phi_1}{\partial x_i}$$

$$\equiv \phi_1 \, \mathrm{grad}\, \phi_2 + \phi_2 \, \mathrm{grad}\, \phi_1$$

(iv) $\mathrm{div}(\phi \mathbf{F}) \equiv \mathbf{e}_i \cdot \dfrac{\partial}{\partial x_i}(\phi \mathbf{F})$

$$\equiv \phi \, \mathbf{e}_i \cdot \frac{\partial \mathbf{F}}{\partial x_i} + \left(\mathbf{e}_i \frac{\partial \phi}{\partial x_i} \right) . \mathbf{F}$$

$$\equiv \phi \, \mathrm{div}\, \mathbf{F} + \mathbf{F} . \, \mathrm{grad}\, \phi$$

The same arguments may be used to prove (v), which is left as an exercise.

PROBLEM 17

Prove (vi) and (vii) using the $\mathbf{e}_i$ notation.

We have

(vi) $\mathbf{F} \times \mathrm{curl}\, \mathbf{G} + \mathbf{G} \times \mathrm{curl}\, \mathbf{F} = \mathbf{F} \times \left(\mathbf{e}_i \times \dfrac{\partial \mathbf{G}}{\partial x_i} \right) + \mathbf{G} \times \left(\mathbf{e}_i \times \dfrac{\partial \mathbf{F}}{\partial x_i} \right)$

$$\equiv \left(\mathbf{F} . \frac{\partial \mathbf{G}}{\partial x_i} \right) \mathbf{e}_i - \left(\mathbf{F} . \mathbf{e}_i \right) \frac{\partial \mathbf{G}}{\partial x_i}$$

$$+ \left(\mathbf{G} . \frac{\partial \mathbf{F}}{\partial x_i} \right) \mathbf{e}_i - \left(\mathbf{G} . \mathbf{e}_i \right) \frac{\partial \mathbf{F}}{\partial x_i}$$

$$\equiv \mathbf{e}_i \frac{\partial}{\partial x_i}(\mathbf{F} . \mathbf{G}) - \left(\mathbf{F} . \mathbf{e}_i \frac{\partial}{\partial x_i} \mathbf{G} \right) - \left(\mathbf{G} . \mathbf{e}_i \frac{\partial}{\partial x_i} \right) \mathbf{F}$$

$$\equiv \mathrm{grad}(\mathbf{F} . \mathbf{G}) - (\mathbf{F} . \nabla)\mathbf{G} - (\mathbf{G} . \nabla)\mathbf{F}$$

proving the result.

(vii) $\text{div}(\mathbf{F} \times \mathbf{G}) \equiv \mathbf{e}_i \cdot \dfrac{\partial}{\partial x_i}(\mathbf{F} \times \mathbf{G}) \equiv \mathbf{e}_i \cdot \left(\dfrac{\partial \mathbf{F}}{\partial x_i} \times \mathbf{G}\right) + \mathbf{e}_i \cdot \left(\mathbf{F} \times \dfrac{\partial \mathbf{G}}{\partial x_i}\right)$

$\equiv \left(\mathbf{e}_i \times \dfrac{\partial \mathbf{F}}{\partial x_i}\right) \cdot \mathbf{G} - \left(\mathbf{e}_i \times \dfrac{\partial \mathbf{G}}{\partial x_i}\right) \cdot \mathbf{F}$

$\equiv \mathbf{G} \cdot \text{curl }\mathbf{F} - \mathbf{F} \cdot \text{curl }\mathbf{G}$

Identity (viii) is proved similarly and is left as an exercise.

EXERCISES ON 9.8

1. Prove the $\mathbf{e}_i$ expressions for grad, div, curl.
2. Prove identities (v) and (viii).
3. Show that if $\mathbf{a}$ is a constant vector, then
   (i) $\text{div}(\mathbf{a} \times \mathbf{r}) = 0$　　　　(ii) $\text{curl}(\mathbf{a} \times \mathbf{r}) = 2\mathbf{a}$
4. Show that

   $\text{curl}(\mathbf{r} \times \text{curl }\mathbf{F}) + (\mathbf{r} \cdot \nabla)\text{curl }\mathbf{F} + 2\,\text{curl }\mathbf{F} \equiv 0$

9.9　The vector operators grad, div and curl in general orthogonal curvilinear coordinates

By now, you are very familiar with the rectangular coordinate system, and have seen a little of cylindrical and spherical coordinates. You know that we try to choose the coordinate system to suit any symmetry properties of the situations for which we need them. In fact, there are a wide range of exotic coordinate systems – *parabolic cylindrical coordinates*, *elliptic cylindrical coordinates*, and so on – all suitable for appropriate purposes. Since we are more familiar with rectangular coordinates, it is customary to express the new 'exotic' systems in terms of transformations from this system. Thus, for example, parabolic cylindrical coordinates (u, v, z) are defined by

$$x = \frac{1}{2}(u^2 - v^2) \qquad y = uv \qquad z = z$$

In fact, we can extend our vector calculus to quite general coordinate systems of a particular type called *curvilinear coordinates*. These, too, can be expressed in terms of transformations from rectangular coordinates. This topic is rather complicated by the plethora of symbols, indices and so on, and so we will only give a brief overview of the ideas, which are actually much simpler than the notation leads one to expect. Essentially, what we have to do is express all the geometrical objects and quantities we have so far met in terms of the new types of coordinates – thus, we need expressions for vectors, arc length, volume element, grad, div and curl in terms of such coordinates.

Let the rectangular coordinates (x, y, z) of any point be expressed as functions of three variables (u_1, u_2, u_3):

$$x = x(u_1, u_2, u_3) \qquad y = y(u_1, u_2, u_3) \qquad z = z(u_1, u_2, u_3)$$

Now, provided the conditions of the implicit function theorem are satisfied we can, in principle if not in practice, solve these equations (invert the transformations) to obtain the u_i in terms of x, y, z:

$$u_1 = u_1(x, y, z) \qquad u_2 = u_2(x, y, z) \qquad u_3 = u_3(x, y, z)$$

So, given a point P with rectangular coordinates (x, y, z) we can use the above equations to represent the point by the coordinates (u_1, u_2, u_3), which are called the *curvilinear coordinates* of P. The previous two sets of equations define a *transformation of coordinates*.

The surfaces $u_1 = c_1$, $u_2 = c_2$, $u_3 = c_3$ are called *coordinate surfaces* and each pair of these intersects in *coordinate curves* or *lines*. In the particularly useful case when the coordinate surfaces intersect at right angles, we say that the curvilinear coordinate system is *orthogonal*.

Now let $\mathbf{r} = x\mathbf{i} + y\mathbf{j} + z\mathbf{k}$ be the position vector of a point P. Using the coordinate transformations, we can express this in terms of coordinates u_1, u_2, u_3, as $\mathbf{r} = \mathbf{r}(u_1, u_2, u_3)$. Consider the u_1-curve at P. Along the u_1-curve, u_2 and u_3 are constant. The quantity $\partial\mathbf{r}/\partial u_1$ is tangent to the u_1-curve. It is in fact a *tangent vector* to the u_1- curve. We now know enough not to take this for granted, of course – it needs to be verified that it transforms like a vector under appropriate coordinate transformations, but we will assume that this is the case. A unit tangent vector in this direction is therefore

$$\mathbf{e}_1 = \frac{\partial\mathbf{r}}{\partial u_1} \bigg/ \left| \frac{\partial\mathbf{r}}{\partial u_1} \right|$$

We write

$$\frac{\partial\mathbf{r}}{\partial u_1} = h_1\mathbf{e}_1$$

where $h_1 = |\partial\mathbf{r}/\partial u_1|$. Similarly, for the other coordinate curves we can define unit tangent vectors $\mathbf{e}_2$, $\mathbf{e}_3$ such that

$$\frac{\partial\mathbf{r}}{\partial u_2} = h_2\mathbf{e}_2 \qquad \frac{\partial\mathbf{r}}{\partial u_3} = h_3\mathbf{e}_3$$

where

$$h_2 = \left| \frac{\partial\mathbf{r}}{\partial u_2} \right| \qquad h_3 = \left| \frac{\partial\mathbf{r}}{\partial u_3} \right|$$

The quantities h_i are called *scale factors*. The unit 'basis vector' $\mathbf{e}_i$ is in the direction of increasing u_i.

∇u_i is normal to the surface $u_i = c_i$ at the point P and so the vectors

$$\mathbf{E}_i = \frac{\nabla u_i}{|\nabla u_i|}$$

are unit vectors normal to the u_i-surface. So, at each point of a curvilinear coordinate system there are two important sets of unit vectors: $\mathbf{e}_1$, $\mathbf{e}_2$, $\mathbf{e}_3$, tangent to the coordinate curves; and $\mathbf{E}_1$, $\mathbf{E}_2$, $\mathbf{E}_3$, normal to the coordinate surfaces. These sets are identical if and only if the curvilinear coordinate system is *orthogonal*. They are analogous to the vectors $\mathbf{i}$, $\mathbf{j}$, $\mathbf{k}$ but in general their direction may vary from point to point.

Any vector **a** can be expressed either in terms of the $\mathbf{e}_i$-*basis* or the $\mathbf{E}_i$-*basis*:

$$\mathbf{a} = a_1\mathbf{e}_1 + a_2\mathbf{e}_2 + a_3\mathbf{e}_3 = A_1\mathbf{E}_1 + A_2\mathbf{E}_2 + A_3\mathbf{E}_3$$

As noted above, the sets of basis vectors and hence the components are the same for orthogonal systems. Since we only consider such systems, we will simply use the $\mathbf{e}_i$ system throughout. However, for more advanced vector and tensor calculus, in which transformations between arbitrary coordinate systems are involved, the distinction between the two sets of basis vectors and the two sets of components becomes important. In such cases the a_i are called *contravariant components* and the A_i are *covariant components*. One reason for the convenience of rectangular coordinates is that contravariant and covariant components are the same.

It is easy to obtain expressions for arc length and volume element in terms of orthogonal curvilinear coordinates.

PROBLEM 18

Show that for orthogonal systems
(i) $ds^2 = h_1^2\,du_1^2 + h_2^2\,du_2^2 + h_3^2\,du_3^2$
(ii) $d\tau = h_1h_2h_3du_1du_2du_3$

(i) In general, we have

$$dr = \frac{\partial \mathbf{r}}{\partial u_1}du_1 + \frac{\partial \mathbf{r}}{\partial u_2}du_2 + \frac{\partial \mathbf{r}}{\partial u_3}du_3$$

$$= h_1du_1\mathbf{e}_1 + h_2du_2\mathbf{e}_2 + h_3du_3\mathbf{e}_3$$

so

$$ds^2 = d\mathbf{r}\cdot d\mathbf{r} = h_1^2\,du_1^2 + h_2^2\,du_2^2 + h_3\,du_3^2$$

on using $\mathbf{e}_i \cdot \mathbf{e}_j = \delta_{ij}$ for orthogonal systems.

Now consider a volume element whose sides are elemental portions of the coordinate curves, as shown in Fig. 9.1.

Along a u_1-curve, u_2 and u_3 are constants and so the element of u_1-curve is $d\mathbf{r} = h_1du_1\mathbf{e}_1$, giving an element of arc length $ds_1 = h_1du_1$ at a point P. The volume element consists of a 'box' with sides represented by vectors $h_1du_1\mathbf{e}_1$, $h_2du_2\mathbf{e}_2$, $h_3du_3\mathbf{e}_3$. The volume of such a box is

$$d\tau = |(h_1du_1\mathbf{e}_1)\cdot(h_2du_2\mathbf{e}_2)\times(h_3du_3\mathbf{e}_3)|$$

$$= h_1h_2h_3du_1du_2du_3$$

on using $|\mathbf{e}_1 \cdot \mathbf{e}_2 \times \mathbf{e}_3| = 1$.

PROBLEM 19

Find the square of the element of arc length ds^2, and the volume element, $d\tau$, in (i) cylindrical coordinates and (ii) spherical coordinates (cf. Exercise 7.9.2).

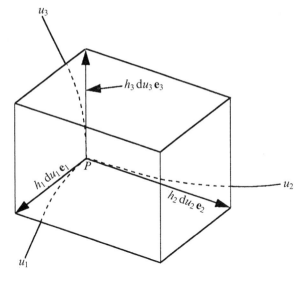

Fig. 9.1

(i) You may be tempted to emulate the general approach given above – that is, find the scale factors and substitute into the results obtained in Problem 18. This is lengthy, however. Instead, we can use the scalar nature of ds^2 as follows. In terms of cylindrical coordinates (r, θ, z) we have

$$x = r \cos \theta \qquad y = r \sin \theta \qquad z = z$$

So

$$dx = -r(\sin \theta)d\theta + (\cos \theta)dr \qquad dy = r(\cos \theta)d\theta + (\sin \theta)dr \qquad dz = dz$$

and therefore

$$ds^2 = dx^2 + dy^2 + dz^2$$
$$= dr^2 + r^2 d\theta^2 + dz^2 \equiv h_1 dr^2 + h_2 d\theta^2 + h_3 dz^2$$

on substituting and simplifying. So the scale factors are $h_1 = 1$, $h_2 = r$, $h_3 = 1$. The volume element is therefore

$$d\tau = h_1 h_2 h_3 du_1 du_2 du_3 = r\,dr\,d\theta\,dz$$

(ii) In spherical coordinates (ρ, θ, ϕ) we have

$$x = \rho \sin \phi \cos \theta \qquad y = \rho \sin \phi \sin \theta \qquad z = \rho \cos \phi$$

So

$$dx = \sin \phi \cos \theta\,d\rho - \rho \sin \phi \sin \theta\,d\theta + \rho \cos \phi \cos \theta\,d\phi$$
$$dy = \sin \phi \sin \theta\,d\rho + \rho \sin \phi \cos \theta\,d\theta + \rho \cos \phi \sin \theta\,d\phi$$
$$dz = \cos \phi\,d\rho \qquad\qquad\qquad\qquad - \rho \sin \phi\,d\phi$$

You can have fun checking that

$$ds^2 = dx^2 + dy^2 + dz^2 = (d\rho)^2 + \rho^2 \sin^2 \phi\,(d\theta)^2 + \rho^2 (d\phi)^2$$

and so the scale factors are $h_1 = 1$, $h_2 = \rho \sin \phi$, $h_3 = \rho$. The volume element is thus

$$d\tau = \rho^2 \sin \phi \, d\rho d\theta d\phi$$

Now consider the case of a general vector field

$$\mathbf{F} = F_1 \mathbf{e}_1 + F_2 \mathbf{e}_2 + F_3 \mathbf{e}_3$$

Note that we must avoid use of the form (F_1, F_2, F_3) in curvilinear coordinates since the basis vectors $\mathbf{e}_i$ are not necessarily constant and, for example,

$$\frac{\partial \mathbf{F}}{\partial u_1} \neq \left(\frac{\partial F_1}{\partial u_1}, \frac{\partial F_2}{\partial u_1}, \frac{\partial F_3}{\partial u_1} \right)$$

in general.

To obtain an expression for $\nabla \phi$ in orthogonal curvilinear coordinates, let

$$\nabla \phi = f_1 \mathbf{e}_1 + f_2 \mathbf{e}_2 + f_3 \mathbf{e}_3$$

where the f_i are to be determined.

From

$$dr = \frac{\partial \mathbf{r}}{\partial u_1} du_1 + \frac{\partial \mathbf{r}}{\partial u_2} du_2 + \frac{\partial \mathbf{r}}{\partial u_3} du_3$$

$$= h_1 \mathbf{e}_1 du_1 + h_2 \mathbf{e}_2 du_2 + h_3 \mathbf{e}_3 du_3$$

we have

$$d\phi = \nabla \phi \cdot dr = h_1 f_1 du_1 + h_2 f_2 du_2 + h_3 f_3 du_3$$

But also we have

$$d\phi = \frac{\partial \phi}{\partial u_1} du_1 + \frac{\partial \phi}{\partial u_2} du_2 + \frac{\partial \phi}{\partial u_3} du_3$$

So equating coefficients of du_i

$$f_i = \frac{1}{h_i} \frac{\partial \phi}{\partial u_i}$$

Hence, we have

$$\nabla \phi = \frac{\mathbf{e}_1}{h_1} \frac{\partial \phi}{\partial u_1} + \frac{\mathbf{e}_2}{h_2} \frac{\partial \phi}{\partial u_2} + \frac{\mathbf{e}_3}{h_3} \frac{\partial \phi}{\partial u_3}$$

for the gradient in orthogonal curvilinear coordinates.

If we put $\phi = u_i$ in the last result we get

$$\mathbf{e}_i = h_i \nabla u_i$$

As the $\mathbf{e}_i$ form a right-handed orthonormal set, we can write, for example,

$$\mathbf{e}_1 = \mathbf{e}_2 \times \mathbf{e}_3 = h_2 h_3 \nabla u_2 \times \nabla u_3$$

So it follows that, for the divergence,

$$\nabla \cdot (F_1 \mathbf{e}_1) = \nabla \cdot (h_2 h_3 \nabla u_2 \times \nabla u_3)$$

$$= (\nabla u_2 \times \nabla u_3) \cdot \nabla (h_2 h_3 F_1)$$

$$= \frac{\mathbf{e}_1}{h_2 h_3} \cdot \nabla (h_2 h_3 F_1)$$

$$= \frac{1}{h_1 h_2 h_3} \frac{\partial}{\partial u_1} (h_2 h_3 F_1)$$

Similarly for $\nabla \cdot (F_2 e_2)$, $\nabla \cdot (F_3 e_3)$, and we can combine the results to get

$$\text{div } \mathbf{F} = \frac{1}{h_1 h_2 h_3} \left[\frac{\partial}{\partial u_1}(h_2 h_3 F_1) + \frac{\partial}{\partial u_2}(h_1 h_3 F_2) + \frac{\partial}{\partial u_3}(h_1 h_2 F_3) \right]$$

for the divergence in orthogonal curvilinear coordinates.

For the curl, we have

$$\nabla \times (F_1 e_1) = \nabla \times (h_1 F_1 \nabla u_1)$$

$$= \nabla(h_1 F_1) \times \nabla u_1$$

$$= \frac{e_2}{h_1 h_3} \frac{\partial}{\partial u_3}(h_1 F_1) - \frac{e_3}{h_1 h_2} \frac{\partial}{\partial u_2}(h_1 F_1)$$

$$= \frac{1}{h_1 h_2 h_3} \begin{vmatrix} h_1 e_1 & h_2 e_2 & h_3 e_3 \\ \dfrac{\partial}{\partial u_1} & \dfrac{\partial}{\partial u_2} & \dfrac{\partial}{\partial u_3} \\ h_1 F_1 & 0 & 0 \end{vmatrix}$$

We get similar results for $\nabla(F_2 e_2)$, $\nabla(F_3\, e_3)$ and combining these gives

$$\text{curl } \mathbf{F} = \frac{1}{h_1 h_2 h_3} \begin{vmatrix} h_1 e_1 & h_2 e_2 & h_3 e_3 \\ \dfrac{\partial}{\partial u_1} & \dfrac{\partial}{\partial u_2} & \dfrac{\partial}{\partial u_3} \\ h_1 F_1 & h_2 F_2 & h_3 F_3 \end{vmatrix}$$

for the curl in orthogonal curvilinear coordinates.

EXERCISES ON 9.9

1. Fill in the details of the derivations of the expressions for div and curl in orthogonal curvilinear coordinates.
2. Show that

$$\nabla^2 = \frac{1}{h_1 h_2 h_3} \left\{ \frac{\partial}{\partial u_1}\left(\frac{h_2 h_3}{h_1} \frac{\partial}{\partial u_1} \right) + \frac{\partial}{\partial u_2}\left(\frac{h_3 h_1}{h_2} \frac{\partial}{\partial u_2} \right) + \frac{\partial}{\partial u_3}\left(\frac{h_1 h_2}{h_3} \frac{\partial}{\partial u_3} \right) \right\}$$

3. Obtain the expressions for grad, div, curl and ∇^2 in (i) cylindrical and (ii) spherical coordinates.

FURTHER EXERCISES

1. Find the derivative of the following functions at the point given in the direction of the vectors given:

   (i) $f(x, y) = 2x + y^2$, $P_0(0, 1)$, $\mathbf{u} = \dfrac{1}{\sqrt{2}}\mathbf{i} + \dfrac{1}{\sqrt{2}}\mathbf{j}$

   (ii) $f(x, y) = e^{xy}\cos(x + y)$, $P_0(2, 1)$, $\mathbf{a} = 2\mathbf{i} + 3\mathbf{j}$

   (iii) $f(x, y, z) = xyz\, e^{xyz}$, $P_0(1, 0, 1)$, $\mathbf{a} = \mathbf{i} + \mathbf{j} + \mathbf{k}$

   (iv) $f(x, y, z) = \dfrac{(x + y + z)}{\sqrt{x^2 + y^2 + z^2}}$, $P_0(-1, 1, -1)$, $\mathbf{a} = 2\mathbf{i} - 3\mathbf{j} + \mathbf{k}$

   (v) $f(x, y, z) = 4e^{2x-y+z}$ at the point $(1, 1, -1)$ in a direction towards the point $(-3, 5, 6)$.

2. Evaluate ∇f for the following functions:
   (i) $f(x, y, z) = x^2 + 2y^2 + 3z^2$ (ii) $f(x, y, z) = xyz \sin(x + y + z)$

   (iii) $f(x, y, z) = e^{xyz} \ln(x + y + z)$ (iv) $f(x, y, z) = \dfrac{1}{\sqrt{x^2 + y^2 + z^2}}$

   (v) $f(x, y, z) = 3xy^2 + 2x^2z + 4xyz$ (vi) $f(x, y, z) = \cos(xy) + \sin(yz)$

3. Find the direction in which the following functions (a) increase, (b) decrease most rapidly, at the given points P_0:
   (i) $f(x, y) = x^2 + 2xy + 3y^2$ $P_0(0, 1)$
   (ii) $f(x, y, z) = xy^2 + 2 \cos(x + y + z)$ $P_0(1, 1, 2)$
   (iii) $f(x, y, z) = \ln(xy) + \ln(yz) + \ln(xz)$ $P_0(1, 1, 1)$

4. Estimate the change in $f(x, y, z) = \ln\sqrt{x^2 + y^2 + z^2}$ in moving from $P_0(3, 4, 12)$ a distance of $ds = 0.1$ units in the direction of $3\mathbf{i} + 6\mathbf{j} - 2\mathbf{k}$.

5. Find the tangent plane and the normal lines to the following surfaces at the given points.
   (i) $x^2 + y^2 + z^2 = 2$ $P_0(1, 0, 1)$
   (ii) $2x - y^2 = 0$ $P_0(2, 2, 0)$

   (iii) $x^2 + \dfrac{y^2}{4} + \dfrac{z^2}{9} = 3$ $P_0(1, 2, 3)$

   (iv) $xyz = 4$ $P_0(1, 2, 2)$
   (v) $z = x^2y$ $P_0(2, 1, 4)$
   (vi) $z = e^{3y} \sin 3x$ $P_0(\pi/6, 0, 1)$

6. If $\mathbf{r} = x\mathbf{i} + y\mathbf{j} + z\mathbf{k}$ find (i) $\nabla |\mathbf{r}|^3$ and (ii) $\nabla f(\mathbf{r})$.

7. Find $\phi(\mathbf{r})$ such that $\nabla\phi = \mathbf{r}/r^5$ and $\phi(2) = 0$.

8. If $\nabla\phi = 2xyz^3\mathbf{i} + x^2z^3\mathbf{j} + 3x^2yz^2\mathbf{k}$, find $\phi(x, y, z)$ if $\phi(1, -2, 2) = 4$.

9. If F is a differentiable function of x, y, z, t, where x, y, z are differentiable functions of t, prove that

$$\frac{dF}{dt} = \frac{\partial F}{\partial t} + \nabla F \cdot \frac{d\mathbf{r}}{dt}$$

10. If $\mathbf{A}$ is a constant vector, prove $\nabla(\mathbf{r} \cdot \mathbf{A}) = \mathbf{A}$.

11. Find the divergence and curl of the following vectors:
    (i) $\mathbf{F} = -\omega y\mathbf{i} + \omega x\mathbf{j}$ (ii) $\mathbf{F} = x\mathbf{i} + y\mathbf{j}$
    (iii) $\mathbf{F} = xyz\mathbf{i} + xzj + z\mathbf{k}$ (iv) $\mathbf{F} = 3xyz^2\mathbf{i} + 2xy^3\mathbf{j} - x^2yz\mathbf{k}$
    (v) $\mathbf{F} = 3x^2y\mathbf{i} + (x^3 + y^3)\mathbf{j}$ (vi) $\mathbf{F} = (xy + z^2)\mathbf{i} + x^2\mathbf{j} + (xz - 2)\mathbf{k}$
    (vii) $\mathbf{F} = f(r)\hat{\mathbf{r}}$, where $\hat{\mathbf{r}}$ is the unit radius vector.
    (viii) $\mathbf{F} = xy\mathbf{i} + (z^2 - 2y)\mathbf{j} + (\cos yz)\mathbf{k}$
    (ix) $\mathbf{F} = 2x^2z\mathbf{i} - xy^2z\mathbf{j} + 3yz^2\mathbf{k}$

12. If $\mathbf{F} = xy\mathbf{i} + yz\mathbf{j} + zx\mathbf{k}$ and $\phi = x^2 + 2y$, find
    (i) $\nabla \cdot \mathbf{F}$ (ii) $\nabla\phi$ (iii) $\nabla \cdot (\phi\mathbf{F})$ (iv) $\nabla \cdot (\nabla\phi)$
    (v) $\nabla \times \mathbf{F}$ (vi) $\mathbf{F} \times (\nabla\phi)$ (vii) $(\mathbf{F} \times \nabla)\phi$ at the point $(1, -1, 1)$.

13. Evaluate $\nabla^2 \phi$ for

    (i) $\phi = x + y + z$ (ii) $\phi = xyz$ (iii) $\phi = 3x^2 y - 2x + x^2 y^3 z$

    (iv) $\phi = e^{-x} \cos y \sin z$ (v) $\phi = \dfrac{1}{r}$ (vi) $\phi = \ln r$

    (vii) $\phi = r^n$

14. Prove that

$$\nabla^2 f(r) = \frac{d^2 f}{dr^2} + \frac{2}{r}\frac{df}{dr}$$

  and find $f(r)$ such that $\nabla^2 f(r) = 0$.

15. Prove that

    (i) $\nabla \times \left(\dfrac{\mathbf{r}}{r^2}\right) = 0$ (ii) $\nabla \times (\phi \nabla \phi) = 0$

16. Essay and discussion topics:

    (i) The gradient as a normal to a surface.

    (ii) The transformation properties of grad, div, curl and their combinations under rotation of axes.

    (iii) Structure and pattern in the vector operator identities.

    (iv) Applications of grad, div, curl in fluid mechanics and electromagnetism.

    (v) Curvilinear coordinates and Jacobians.

10 • Integration in Vector Fields

10.1 Introduction

Many of the applications of vector calculus arise in mathematical physics – fluid and solid mechanics, electromagnetism, etc. It is therefore usual to motivate some of the more advanced ideas of vector calculus by starting with physical discussions. I am not keen to do that since it assumes significant physical knowledge from the reader – and it detracts from the essential fact that we are talking about *mathematics*, not *physics*. However, on this occasion the physical motivation is so informative, and so fundamental, that it is worth the risk to use it. I apologize if it involves you in a bit of work!

The *work* done in moving a force through a distance is defined as force × distance in elementary mechanics. This is good enough if the force is constant and in the direction of the motion. If the force is constant but at an angle to the direction of motion, then we use the vector form, $\mathbf{F} \cdot \mathbf{d}$, for work done. This is in fact one of the reasons why the scalar product is defined as it is. But what happens if $\mathbf{F}$ is a vector field (like gravity, for example), varying in magnitude and direction from point to point, and the motion is between two points on a curve?

Suppose the force field is given by

$$\mathbf{F} = F_1(x, y, z)\mathbf{i} + F_2(x, y, z)\mathbf{j} + F_3(x, y, z)\mathbf{k}$$

and suppose it acts over a continuous curve

$$C\colon \mathbf{r}(t) = x(t)\mathbf{i} + y(t)\mathbf{j} + z(t)\mathbf{k}$$

from $t = a$ to $t = b$. Then the work done in moving along a small element of the curve $\mathbf{T}\,\mathrm{d}s$, where $\mathbf{T}$ is the unit tangent and $\mathrm{d}s$ is the arc length of the element, is $\mathbf{F} \cdot \mathbf{T}\,\mathrm{d}s = \mathbf{F}(\mathbf{r}(t)) \cdot \mathbf{r}'(t)\,\mathrm{d}t$, since we can assume that the force $\mathbf{F}$ is constant over this element. So the total work done by $\mathbf{F}$ in moving along the curve from $t = a$ to $t = b$ is

$$W = \int_a^b \mathbf{F}(\mathbf{r}(t)) \cdot \mathbf{r}'(t)\,\mathrm{d}t$$

Note that in general the work done will depend on the curve traced between a and b. Vectorially, the integral expression for W is a scalar quantity. It is called the *tangential line integral of the vector field* $\mathbf{F}$ *along the curve C.*

In fluid dynamics and electromagnetism we often have occasion to consider the 'flux' of a vector field quantity across a surface. This leads us to another important type of integral of a vector field. Thus, suppose $\mathbf{F}$ is a vector field, and let S be a surface with outward unit normal represented by $\mathbf{n}$. Then $\mathbf{F} \cdot \mathbf{n}$ is the component of

F normal to the surface at a point with normal **n**. We say that the *flux* of **F** across an element of area $d\sigma$ of the surface S, with normal **n**, is **F** . **n** $d\sigma$. The flux of **F** across the surface S is then given by the *surface integral*

$$\iint_S \mathbf{F} . \mathbf{n} \, d\sigma$$

Having defined line and surface integrals we will be in a position to discuss three great theorems of integration in vector calculus – *Green's theorem, Gauss's divergence theorem* and *Stokes's theorem*. These theorems look complicated at first sight, but actually they all have a very simple foundation. They essentially relate an n-tuple integration to an $(n-1)$-tuple integration, by the general process

an integral over a region R = a related integral over the boundary of R

For example, we might have

surface integral over R = line integral around the boundary of R
(double)   (single)

Another viewpoint, related to this, is that they come about by a sort of generalized integration by parts.

Note that in this chapter I am going to use R to denote a general region, rather than the Ω used in Chapter 7. I used Ω in Chapter 7 because I wanted to reserve R for rectangular regions in the discussion of multiple integrals – there is no need for the distinction now.

10.2 Line integrals

Let

$$\mathbf{F} = F_1(x, y, z)\mathbf{i} + F_2(x, y, z)\mathbf{j} + F_3(x, y, z)\mathbf{k}$$

be a vector field continuous on a smooth curve

$$C: \mathbf{r}(t) = x(t)\mathbf{i} + y(t)\mathbf{j} + z(t)\mathbf{k} \qquad t \in [a, b]$$

Then the *line integral* of **F** over C is the scalar

$$\int_C \mathbf{F}(\mathbf{r}) . \, d\mathbf{r} = \int_a^b \mathbf{F}(\mathbf{r}(t)) . \, \mathbf{r}'(t) \, dt$$

The parametrization of the curve is important. We can have any number of different parametrizations of a given smooth curve, and it is necessary to be sure that different parametrizations do not lead to different values for the line integral. This will be the case provided the different parametrizations describe the same smooth curve and traverse the curve exactly once as t ranges over its values. Converting from one such parametrization to another is referred to as a *proper reparametrization*.

PROBLEM 1

Evaluate the line integral of the vector field

$$\mathbf{F} = x\mathbf{i} + y\mathbf{j} + z\mathbf{k}$$

over the curve

$$C : \mathbf{r}(t) = t^3\mathbf{i} + t^2\mathbf{j} + t\mathbf{k} \qquad t \in [-1, 1]$$

Along the curve we have

$$x = t^3 \quad y = t^2 \quad z = t$$

So on the curve

$$\mathbf{F} = t^3\mathbf{i} + t^2\mathbf{j} + t\mathbf{k}$$

and

$$\mathbf{r}' = 3t^2\mathbf{i} + 2t\mathbf{j} + \mathbf{k}$$

and therefore

$$\mathbf{F} \cdot \mathbf{r}' = 3t^5 + 2t^3 + t$$

The line integral is thus

$$\int_C \mathbf{F} \cdot d\mathbf{r} = \int_{-1}^{1} \mathbf{F} \cdot \mathbf{r}' \, dt = \int_{-1}^{1} (3t^5 + 2t^3 + t) \, dt = 0$$

Note that there are a number of useful alternative forms for the line integral:

$$\int_C \mathbf{F} \cdot d\mathbf{r} = \int_a^b \mathbf{F}(\mathbf{r}(t)) \cdot \mathbf{r}'(t) \, dt = \int_a^b \mathbf{F} \cdot \mathbf{T} \, ds$$

$$= \int_a^b \left(F_1\frac{dx(t)}{dt} + F_2\frac{dy(t)}{dt} + F_3\frac{dz(t)}{dt} \right) dt$$

$$= \int_{\mathbf{r}_1}^{\mathbf{r}_2} (F_1 \, dx + F_2 \, dy + F_3 \, dz)$$

where $\mathbf{r}_1 = \mathbf{r}(a)$ is the initial point of C and $\mathbf{r}_2 = \mathbf{r}(b)$ is the final point. An object of the form $F_1 \, dx + F_2 \, dy + F_3 \, dz$ is called a *differential form*.

PROBLEM 2

Show that the following parametrizations represent the same curve.

$$\mathbf{r} = (t + 3)\mathbf{i} + (t^2 - t + 2)\mathbf{j} \qquad t \in (-\infty, \infty)$$
$$\mathbf{r} = u\mathbf{i} + (u^2 - 7u + 14)\mathbf{j} \qquad u \in (-\infty, \infty)$$

Evaluate

$$\int_{(0,14)}^{(1,8)} (2x^2\mathbf{i} - y\mathbf{j}) \cdot d\mathbf{r}$$

for each parametrization and verify that the result is the same in each case.

That the two parametrizations represent the same curve is readily checked by noting that we can transform between the two by the transformation $u = t + 3$. Or, notice that eliminating t and u from the parametrizations

$$x = t + 3 \qquad y = t^2 - t + 2$$
$$x = u \qquad y = u^2 - 7u + 14$$

produces the equation

$$y = x^2 - 7x + 14$$

in each case.

For the t-parametrization we have

$$\int_{(0,14)}^{(1,8)} (2x^2\mathbf{i} - y\mathbf{j}) \cdot d\mathbf{r} = \int_{-3}^{-2} (2(t+3)^2\mathbf{i} - (t^2 - t + 2)\mathbf{j}) \cdot (\mathbf{i} + (2t - 1)\mathbf{j}) \, dt$$

$$= \int_{-3}^{-2} (-2t^3 + 5t^2 + 7t + 20) \, dt = \frac{200}{3}$$

Similarly, for the u-parametrization we have

$$\int_{(0,14)}^{(1,8)} (2x^2\mathbf{i} - y\mathbf{j}) \cdot d\mathbf{r} = \int_{0}^{1} (2u^2\mathbf{i} - (u^2 - 7u + 14)\mathbf{j}) \cdot (\mathbf{i} + (2u - 7)\mathbf{j}) \, du$$

$$= \int_{0}^{1} (-2u^3 + 23u^2 - 77u + 98) \, du = \frac{200}{3}$$

as before.

You can see that all the reparametrization really amounts to is a change of dummy variable in the final integration, from t to u.

A thought might strike you, particularly if you know some contour integration. Do we always need to parametrize the curve? In ordinary integration of functions of a single variable 'parametrization' is essentially a change of variable. It is not always *necessary* – sometimes we can integrate the function directly. Also, more than one substitution might perform the same task – evaluate the same integral. Look at the 'differential form' expression for the line integral, and consider the case of the vector field $\mathbf{F} = 3x^2\mathbf{i} + y\mathbf{j} + 2z^3\mathbf{k}$. Forget the curve we integrate along – just take the endpoints to be $\mathbf{r}_1 = (x_1, y_1, z_1)$ and $\mathbf{r}_2 = (x_2, y_2, z_2)$. Then the line integral becomes

$$\int_{\mathbf{r}_1}^{\mathbf{r}_2}(3x^2\,dx + y\,dy + 2z^3\,dz) = \int_{\mathbf{r}_1}^{\mathbf{r}_2}3x^2\,dx + \int_{\mathbf{r}_1}^{\mathbf{r}_2}y\,dy + \int_{\mathbf{r}_1}^{\mathbf{r}_2}2z^3\,dz$$

$$= [x^3]_{\mathbf{r}_1}^{\mathbf{r}_2} + \left[\frac{y^2}{2}\right]_{\mathbf{r}_1}^{\mathbf{r}_2} + \left[\frac{z^4}{2}\right]_{\mathbf{r}_1}^{\mathbf{r}_2} = x_2^3 - x_1^3 + \frac{y_2^2 - y_1^2}{2} + \frac{z_2^4 - z_1^4}{2}$$

There was no need for parametrization in this case because we could indeed integrate directly. And it does not matter what the curve between the points is – the result depends only on the endpoints.

So, was there something special about the vector field in this case, that allowed us to integrate directly?

PROBLEM 3

> Have a look at some elementary calculus books and remind yourself about the fundamental theorem of integral calculus.

The following statement of the theorem will do (Pearson, 1996):
Suppose $f(x)$ is defined and continuous at all x on a closed interval $a \le x \le b$. Then

$$\frac{d}{dx}\int_a^x f(t)\,dt = f(x) \qquad (a \le x \le b)$$

That is, the function $\int_a^x f(t)\,dt$ is an antiderivative of $f(x)$ on the prescribed interval.

An alternative way of looking at this is that if $f(x) = F'(x)$, then

$$\int_a^b f(x)\,dx = \int_a^b F'(x)\,dx = F(b) - F(a)$$

In this case the key thing is that when integrating a derivative the integration can be performed immediately, and the result depends only on the values of the differentiated function at the limits of integration. Now in vector calculus, it is the *gradient* which corresponds to the derivative. This leads us to:

● *The fundamental theorem for line integrals*

Let $C: \mathbf{r} = \mathbf{r}(t)$, $t \in [a, b]$, be a piecewise smooth curve with initial point $\mathbf{a} = \mathbf{r}(a)$ and final point $\mathbf{b} = \mathbf{r}(b)$. If the scalar field ϕ is continuously differentiable on an open set that contains the curve C, then

$$\int_C \nabla\,\phi(\mathbf{r}) \cdot d\mathbf{r} = \phi(\mathbf{b}) - \phi(\mathbf{a})$$

PROOF
Since C is smooth we have

$$\int_C \nabla\phi(\mathbf{r}) \cdot d\mathbf{r} = \int_a^b \nabla\phi(\mathbf{r}(t)) \cdot \mathbf{r}'(t) \, dt$$

$$= \int_a^b \frac{d}{dt}[\phi(\mathbf{r}(t))] \, dt$$

by the chain rule

$$= \phi(\mathbf{r}(t)) \Big|_a^b = \phi(\mathbf{r}(b)) - \phi(\mathbf{r}(a)) = \phi(\mathbf{b}) - \phi(\mathbf{a})$$

Note that in this case, when the vector field $\mathbf{F} = \nabla\phi(\mathbf{r})$ is a *gradient field* then its line integral depends only on the endpoints of the path and not on the path itself.

An important corollary is that if the curve C is *closed* so that $\mathbf{a} = \mathbf{b}$, then

$$\int_C \nabla\phi(\mathbf{r}) \cdot d\mathbf{r} = 0$$

It is necessary to mention a proviso here about the sorts of region in which the closed curve C resides. Crudely speaking, we have to beware of regions which contain 'holes' – annular regions, for example. Integration of a gradient field around a closed curve which encircles a hole (e.g. around the annulus) is not necessarily zero. Also, the theorems we are about to discuss in this chapter need care when dealing with regions with 'holes'. We do not have the space to go into detail here, but it is possible to deal with such regions and extend the theorems to them with appropriate modifications. The characterization of such regions is achieved by considering simple closed curves within them. If every such curve within the region can be shrunk to a point without leaving the region, then the region is called *simply-connected* – it has no holes. If, on the other hand, the region does contain a curve which cannot be so contracted without leaving the region, then the region is said to be *multiply-connected*.

It is interesting to return briefly to how we first met the line integral in Section 10.1, as an expression for the work done by a force moving along a curve. We might expect that the work done by such a force in moving between two points depends on the curve taken between the two points. For quite general forces this would indeed be the case, but we have just seen that there are vector fields for which the line integral is independent of the path taken between two points. A force described by such a field is said to be a *conservative force*. From the result proved above, we see that such forces are characterized by gradient fields. The scalar field of which the force field is a gradient is particularly important, and leads to the following definition.

● *Potential energy functions* ─────────────────────────────

If $\mathbf{F}$ is a conservative field then functions U for which $\mathbf{F} = -\nabla U$ are called *potential energy functions* for $\mathbf{F}$. Note that the sign is chosen so that U conforms with the mechanical definition of potential energy. Also, because the definition of U is via a gradient, U is always arbitrary up to an additive quantity whose gradient is zero – hence we refer to *functions*.

Conservative fields are so called because the *total mechanical energy* of any particle moving in such a field is *conserved*, i.e. remains constant. To see this we start from the equation of motion:

$$m\mathbf{r}'' = \mathbf{F}$$

We can take the scalar product of both sides with the velocity $\mathbf{r}'$ to get

$$m\mathbf{r}' \cdot \mathbf{r}'' = \frac{m}{2}\frac{d}{dt}(\mathbf{r}' \cdot \mathbf{r}') = \mathbf{F} \cdot \mathbf{r}'$$

So for a field derived from a potential, $\mathbf{F} = -\nabla U$, we have

$$\frac{d}{dt}\left(\frac{m}{2}\mathbf{r}'^2\right) = -\nabla U \cdot \mathbf{r}'$$

$$= -\frac{d}{dt}(U(\mathbf{r}))$$

by the chain rule. Hence

$$\frac{d}{dt}\left(\frac{m}{2}\mathbf{r}'^2 + U(\mathbf{r})\right) = 0$$

or

$$\frac{m}{2}\mathbf{r}'^2 + U(\mathbf{r}) = \text{constant}$$

i.e.

kinetic energy + potential energy = constant

Notice that the fact that we can write $\mathbf{F} = -\nabla U$ is crucial in this derivation.

Often, we know when a vector field is a gradient field, and the fundamental theorem can be easily applied. If, on the other hand, we have to test the field to see if it is a gradient, and have to find what it is the gradient of, then things are more problematical. Clearly, it will be useful to have a nice simple test for whether a given field is a gradient. We will return to this in Section 10.6, but just for now have a look through the vector identities in Section 9.8 and see if they suggest any ideas.

In fluid dynamics the line integral $\int_a^b \mathbf{F} \cdot d\mathbf{r}$ of $\mathbf{F}$ along a curve $\mathbf{r} = x(t)\mathbf{i} + y(t)\mathbf{j} + z(t)\mathbf{k}$ is called the *flow* of $\mathbf{F}$ along the curve. This is because if $\mathbf{F}$ represents the velocity field of a fluid flowing through a region of space, then the line integral gives the fluid's flow along the curve. If the curve is closed, then the flow is called the *circulation* around the curve. On the other hand, the integral

$$\int_C \mathbf{F} \cdot \mathbf{n}\, ds$$

where $\mathbf{n}$ is the unit outward normal to the curve C, and s is the arc length, is called the *flux* of $\mathbf{F}$ across the curve.

1. Calculate the line integral $\int_C F(\mathbf{r}) . d\mathbf{r}$ where $F(\mathbf{r}) = xe^y\mathbf{i} + xy\mathbf{j}$ and C is the rectangle, traversed in the anti-clockwise direction, which connects the points $(0, 0), (1, 0), (1, 2), (0, 2)$.

2. Evaluate the line integral $\int_C \mathbf{F} . d\mathbf{r}$ for
   (i) $\mathbf{F}(\mathbf{r}) = x^2\mathbf{i} + y^2\mathbf{j} + z^2\mathbf{k}$
       C is $\mathbf{r}(t) = (\cos t)\mathbf{i} + (\sin t)\mathbf{j} + t\mathbf{k}$ from $(1, 0, 0)$ to $(0, 1, \pi/2)$
   (ii) $\mathbf{F}(\mathbf{r}) = y^2z^3\mathbf{i} + 2xyz^3\mathbf{j} + 3xy^2z^2\mathbf{k}$
        C is $\mathbf{r}(t) = t^2\mathbf{i} + t^4\mathbf{j} + t^6\mathbf{k}$, from $t = 0$ to $t = 1$

3. Show that the work done on a particle in moving from A to B is equal to the change in kinetic energies at these points whether the force field is conservative or not.

10.3 Green's theorem in a plane

This is a fundamental theorem in multi-variable calculus on which the more physically important divergence theorem and Stokes's theorem are based.

Essentially, Green's theorem (which can take a couple of forms: flux–divergence or circulation–curl) gives us a relationship between a line integral around a closed curve in a plane and a double integral over the region enclosed by that curve. The means by which it is extended to the divergence and Stokes's theorems is to break a surface up into adjoining flat plates (like a geodesic dome).

● *Green's theorem in the plane (circulation–curl form, or tangential form)* ────────────────

Let R be a region bounded by a simple-closed piecewise smooth plane curve C (such a region is called a *Jordan region*). If P and Q are scalar fields which are continuously differentiable on an open set that contains R, then

$$\oint_C P(x, y)\,dx + Q(x, y)\,dy = \iint_R \left[\frac{\partial Q(x, y)}{\partial x} - \frac{\partial P(x, y)}{\partial y} \right] dx\,dy$$

where the line integral on the left is taken over C in the anti-clockwise direction.

Viewed cold, this might look a bit confusing and structureless. However, it may help to note that the result gives the anti-clockwise *circulation* of a vector field around a simple-closed curve as the double integral of the curl of the field over the region enclosed by the curve. This also serves to presage *Stokes's theorem* which is essentially the three-dimensional or vector extension of the circulation–curl form of Green's theorem.

PROOF
For general regions the proof is complicated. We will start with simple regions which are of both type I and type II, as defined in Chapter 7.

If we assume R is of type I then we may deal with the $\partial P/\partial y$ integral as follows, with the notation used in Section 7.4.

$$\iint\limits_{R} -\frac{\partial P(x,\,y)}{\partial y}\,dx\,dy = -\int\limits_{a}^{b}\int\limits_{\phi_1(x)}^{\phi_2(x)}\frac{\partial P(x,\,y)}{\partial y}\,dx\,dy$$

$$= -\int\limits_{a}^{b}[P(x,\,\phi_2(x)) - P(x,\,\phi_1(x))]\,dx \qquad (10.1)$$

on using the fundamental theorem of integral calculus.

Now consider the line integral of $P(x,\,y)$ around the contour C. From the form of a type I region this will consist of the line integral around

$$C_1: \mathbf{r}_1(t) = t\mathbf{i} + \phi_1(t)\mathbf{j} \qquad t \in [a,\,b]$$

followed by $-C_2$ where

$$C_2: \mathbf{r}_2(t) = t\mathbf{i} + \phi_2(t)\mathbf{j} \qquad t \in [a,\,b]$$

and so

$$\oint\limits_{C} P(x,\,y)\,dx = \int\limits_{C_1} P(x,\,y)\,dx - \int\limits_{C_2} P(x,\,y)\,dx$$

$$= \int\limits_{a}^{b} P(t,\,\phi_1(t))\,dt - \int\limits_{a}^{b} P(t,\,\phi_2(t))\,dt$$

$$\equiv \int\limits_{a}^{b} P(x,\,\phi_1(x))\,dt - \int\limits_{a}^{b} P(x,\,\phi_2(x))\,dx$$

Comparing with (10.1), we therefore get

$$\oint\limits_{C} P(x,\,y)\,dx = \iint\limits_{R} -\frac{\partial P(x,\,y)}{\partial y}\,dx\,dy$$

On the assumption that R is a type II region, a similar argument demonstrates that

$$\oint\limits_{C} Q(x,\,y)\,dy = \iint\limits_{R} \frac{\partial Q(x,\,y)}{\partial x}\,dx\,dy$$

So we have proved Green's theorem for a region which is of both type I and type II.

We can extend this to more complicated regions by splitting them up into 'elementary regions', as shown in Fig. 10.1. The surface integrals over the two regions R_1, R_2 may be added, while the line integral round the curve enclosing R_1 and R_2 may be formed by adding the two separate line integrals round R_1 and R_2 – the integrals along the connecting line cancel out because they are traversed in opposite directions. We can apply similar tricks to deal with multiply-connected regions in general.

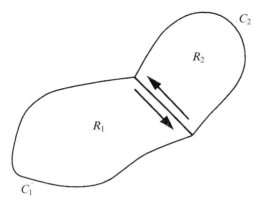

Fig. 10.1

The above *circulation–curl* or *tangential form* of Green's theorem can be written in the vector form:

$$\oint_C \mathbf{F} \cdot \mathbf{T} \, ds = \iint_R (\nabla \times \mathbf{F}) \cdot \mathbf{k} \, dx \, dy$$

where $\mathbf{k}$ is the unit normal to the 'surface' R and $\mathbf{T}$ is the unit tangent vector to the curve C. In this form it suggests Stokes's theorem of Section 10.6 very directly – but the above applies to *plane surfaces*, while Stokes's theorem applies to general curved surfaces.

An alternative form of Green's theorem in the plane, called the *flux–divergence* or *normal form*, may be obtained by applying the tangential form to the vector field $\mathbf{G} = -Q\mathbf{i} + P\mathbf{j}$ to give

$$\oint_C \mathbf{G} \cdot \mathbf{T} \, ds = \oint_C P \, dy - Q \, dx = \iint_R \left(\frac{\partial P}{\partial x} + \frac{\partial Q}{\partial y} \right) dx \, dy$$

Vectorially, this takes the form

$$\oint_C \mathbf{F} \cdot \mathbf{n} \, ds = \iint_R \nabla \cdot \mathbf{F} \, dx \, dy$$

where $\mathbf{n}$ is the unit outward normal to C, and $\mathbf{F} = P\mathbf{i} + Q\mathbf{j}$.

The extension to the three-dimensional *divergence theorem* of Section 10.5 is not quite so transparent here, but it is close enough to explain the alternative 'flux–divergence' terminology. So, this connection to the divergence and Stokes's theorems explains well the immense theoretical importance of Green's theorem, which, you will notice from the proof, originates mainly from the application of the fundamental theorem of integral calculus. However, Green's theorem is also of great practical use. By replacing a line integral by a double integral, or vice versa, it gives us a choice of how to evaluate such objects, and one form is sometimes easier than the other.

PROBLEM 4

Use Green's theorem to evaluate

$$\oint_C (x^2 + 2y)\, dx + (x + 3y^2)\, dy$$

where C is the circle $x^2 + y^2 = 1$.

In this case we take the region R to be the closed disc $0 \leq x^2 + y^2 \leq 1$ and we have $P(x, y) = x^2 + 2y$, $Q(x, y) = x + 3y^2$. So

$$\frac{\partial Q}{\partial x} - \frac{\partial P}{\partial y} = -1$$

and Green's theorem gives

$$\oint_C (x^2 + 2y)\, dx + (x + 3y^2)\, dy = \iint_R -1\, dx\, dy$$

$$= -\text{ area of } R = -\pi$$

This illustrates vividly how useful Green's theorem can be. In this case the line integral could have been evaluated directly using the parametrization

$$x = \cos t \qquad y = \sin t \qquad t \in [0, 2\pi]$$

I will leave this to you as an exercise – which will probably endear you to Green as much as anything else will.

The circulation–curl form of Green's theorem suggests an interesting result related ultimately to the concept of the total differential. If it is the case that

$$\frac{\partial Q}{\partial y} = \frac{\partial P}{\partial x}$$

everywhere on the region R, then it follows from Green's theorem that

$$\oint_C P(x, y)\, dx + Q(x, y)\, dy = 0$$

for every simple-closed piecewise smooth curve C in R. As a consequence of this, the integral

$$\int_{(x_1, y_1)}^{(x_2, y_2)} P(x, y)\, dx + Q(x, y)\, dy$$

must be independent of the path between the points (x_1, y_1), (x_2, y_2), provided it lies wholly within R.

On the other hand, we know from the theory of exact differentials (Cox, 1996) that provided $P(x, y)$ and $Q(x, y)$ have continuous partial derivatives in the region R then a necessary and sufficient condition for $P\, dx + Q\, dy$ to be an exact differential is precisely $\partial Q/\partial y = \partial P/\partial x$. Hence, in this case there exists a function $f(x, y)$ such that

$$P\,dx + Q\,dy \equiv df$$

on R and so for any simple-closed piecewise smooth curve C in R we have

$$\oint_C P\,dx + Q\,dy = \oint_C df = 0$$

This explains the result we deduced from Green's theorem – the equality of $\partial Q/\partial y$ and $\partial P/\partial x$ implies that $P\,dx + Q\,dy$ is a total differential, which will therefore produce zero on integration round a closed curve.

Note that we can apply Green's theorem to multiply-connected regions using the trick illustrated in Fig. 10.2.

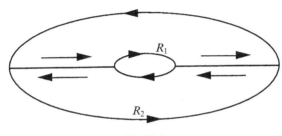

Fig. 10.2

EXERCISES ON 10.3

1. Use Green's theorem to evaluate

    (i) $\displaystyle\oint_C (x - y)\,dx + (y - x)\,dy$ (ii) $\displaystyle\oint_C (y - x)\,dx - (x - y)\,dy$

    where C is the square with vertices $(0, 0)$, $(1, 0)$, $(1, 1)$, $(0, 1)$. Comment on your result for (i).

2. Verify Green's theorem for

$$\oint_C (x - y)\,dx + x\,dy$$

    where C is the unit circle.

10.4 Surface integrals

The usefulness of Green's theorem in a plane immediately prompts us to ask if/how we can extend it to three dimensions. In three dimensions we will have to deal with curved surfaces. We can extend the circulation–curl form of Green's theorem to three dimensions if we can extend the surface integral over the plane to a surface integral over a curved surface. We can do this by regarding the curved surface as constructed from myriads of small flat plates. We can then apply Green's theorem to each flat plate, and combine the results. This is easier said than done – when we extended Green's theorem to non-elementary regions we simply had to combine results over adjacent flat plates in a plane. Now we will have to combine results over adjacent flat plates *at an angle to each other*. I can see you getting worried already – and to be honest this *is* an awkward topic, so we will take it slowly. First, we need to discuss surfaces and their representation more carefully.

We first need a few basic notions about surfaces – analogous to those we needed with curves. The very idea of a surface is not quite so simple as it might seem. For example, you may be aware that a surface may have only one side. Such a surface can be obtained by twisting a strip of paper once and gluing the ends together. This produces a one-sided surface, called a *Möbius strip*. That this surface has only one side can be verified by tracing a continuous line along the strip and confirming that all points can be reached without leaving the strip. In this book we will only consider two-sided surfaces, and will rely on the obvious common-sense view of what such a surface is. In particular, a *closed* surface is one which contains a volume – it divides space into two regions, 'inside' and 'outside', which cannot be connected without crossing the surface. An example of a closed surface would be a spherical shell. A surface is *open* if it does not divide space in this way – there is no 'outside' or 'inside' volume. An example of an open surface is a cap of a sphere. A surface is said to be *smooth* if the unit normal exists and is continuous at all points of the surface. A *simple* or *piecewise smooth* surface is one which is a union of a finite number of smooth surfaces. For example, a cap of a sphere is a smooth surface. An example of a simple-closed surface is the surface of a cube, which is the union of six smooth open surfaces.

If we think of functions of a single variable, $y = f(x)$, and their geometrical representation as curves, then there are three common forms of representation: *explicit*, as $y = f(x)$; *implicit*, as something like $F(x, y) = 0$; and *parametric*, as $x = x(t)$, $y = y(t)$. Similarly, we can represent surfaces explicitly as $z = f(x, y)$, implicitly as $F(x, y, z) = 0$ and parametrically as described below. For now, simply take note that the form of the expressions for *surface area* and *surface integrals* will differ for the different representations. This does not make learning the topic any easier! First, let us look at the parametrization of a surface.

Consider the surface

$$x^2 + y^2 + z^2 = 1$$

which is the surface of a sphere of unit radius, centred on the origin. The trouble with this representation is that the relation between the coordinates renders at least one redundant. Remember what we did in the case of a curve $y = f(x)$ – we introduced a parameter, t, in terms of which x and y could be expressed on the curve itself; indeed, if t is taken as length along the curve then it literally is a coordinate 'along the curve'. The effect of this is to introduce a single *parameter*, corresponding to the single *dimension* described by the curve. I do not need to remind you how useful this was for evaluating and indeed defining such things as line integrals.

So, it would clearly be useful to have some similarly economic parametrization of a surface, in which points (x, y, z) on the surface are described in terms of one or more parameters defined on the surface. In fact, since a surface is two-dimensional we expect to have to use two such parameters, call them u and v. We are going to assume that our surface can be projected uniquely point for point into some region of *xyz*-space. Not all the surfaces that are useful to us necessarily satisfy this condition, but where it breaks down the difficulties it yields can be left to more advanced texts.

So, suppose a surface, *S*, is defined by points with position vectors (cf. $\mathbf{r}(t)$).

$$\mathbf{r}(u, v) = (x(u, v), y(u, v), z(u, v))$$

where *u*, *v* are continuous parameters taking all values in some region *W* of the *uv*-plane, and *x*, *y*, *z* are continuous, single-valued functions of *u*, *v* in *W*. Note that we must be careful to distinguish regions in the *uv*-plane from those in the *xyz*-space in which surfaces described by *u*, *v* are embedded. Hence *W* for the former, *R* for the latter, for example.

By fixing *v* (*u*) and allowing *u* (*v*) to vary we produce a *u*- (*v*-)*coordinate curve* on the surface, and the surface may be regarded as being covered by a mesh of such coordinate curves. In a flat plane the *u*-, *v*-coordinate curves may be the *x*- and *y*-axes for example – straight lines. Now we know how useful the basis vectors **i**, **j** are in a plane – they essentially replace the coordinate curves by vectors. Can we do this on a curved surface?

Well, consider a point *P*(*u*, *v*) on the surface *S*. Through *P* will be two *u*- and *v*-coordinate curves. Denote by $\mathbf{r}_u$, $\mathbf{r}_v$ the tangent vectors to these curves at *P*

$$\frac{\partial \mathbf{r}}{\partial u} = \mathbf{r}_u \qquad \frac{\partial \mathbf{r}}{\partial v} = \mathbf{r}_v$$

(see Fig. 10.3). Note carefully the distinction between the $\partial \mathbf{r}/\partial u$ used here (*u* is a parameter defining a surface) and the $\partial \mathbf{r}/\partial u_1$, for example, used in Section 9.9 (u_1 is an alternative variable in the same space as *x*, *y*, *z*).

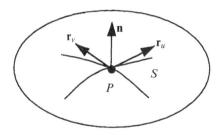

Fig. 10.3

The vector

$$\mathbf{n} = \frac{\mathbf{r}_u \times \mathbf{r}_v}{|\mathbf{r}_u \times \mathbf{r}_v|}$$

is normal to the surface at *P*, and of unit magnitude – it is the *unit normal vector* to *S* at *P*. The direction of **n** is not unique – it may be pointing 'outwards' or 'inwards' from the surface. Clearly these terms need to be defined by some convention and a choice made for **n**. The convention adopted is to define the *positive* normal to a surface bounding a closed region *R* to be that which points out of the region *R*. Once one side of a surface has been labelled as positive, we say the surface is *oriented.* The vector product $\mathbf{r}_u \times \mathbf{r}_v$ is sometimes referred to as the *fundamental vector product for the surface.*

PROBLEM 5

> Obtain the outward unit normal vector to the circular cylinder
>
> $$\mathbf{r} = (a \cos \theta, a \sin \theta, z) \qquad \theta \in [0, 2\pi], z \in (-\infty, \infty)$$
>
> where (a, θ, z) are in cylindrical coordinates.

In this case we have

$$\mathbf{r}_\theta = (-a \sin \theta, a \cos \theta, 0)$$
$$\mathbf{r}_z = (0, 0, 1)$$

We now have to decide whether $\mathbf{r}_\theta \times \mathbf{r}_z$ or $\mathbf{r}_z \times \mathbf{r}_\theta$ gives the outward normal. If you refer back to Fig. 2.4 you will see that with $\mathbf{r}_\theta, \mathbf{r}_z$ pointing in the directions of increasing θ and increasing z respectively, $\mathbf{r}_\theta \times \mathbf{r}_z$ actually points out of the cylinder.

So, for the outward normal we take

$$\mathbf{n} = \frac{\mathbf{r}_\theta \times \mathbf{r}_z}{|\mathbf{r}_\theta \times \mathbf{r}_z|} = (\cos \theta, \sin \theta, 0)$$

Putting $\theta = 0$, for example, gives $\mathbf{n} = (1, 0, 0) = \mathbf{i}$, confirming that the unit normal points out of the cylinder.

We are now in a position to define surface area. Let $P_0(u, v)$ be a point on a surface S, and suppose $P_1(u + du, v)$, $P_2(u, v + dv)$ are neighbouring points on the u-coordinate and v-coordinate curves through P_0, respectively. Suppose these coordinate curves meet at P_3. Then we call the part of S bounded by P_0, P_1, P_2, P_3 a *surface element* of the surface S.

We can, provided it is sufficiently small, approximate the surface element by a parallelogram with area:

$$d\sigma \simeq \left| (\overrightarrow{P_0 P_1} \times \overrightarrow{P_0 P_2}) \right| \simeq |\mathbf{r}_u \times \mathbf{r}_v| \, du \, dv$$

since

$$\overrightarrow{P_0 P_1} \simeq \frac{\partial \mathbf{r}}{\partial u} du \qquad \overrightarrow{P_0 P_2} \simeq \frac{\partial \mathbf{r}}{\partial v} dv$$

We use this heuristic argument to motivate the definition of the *surface area of S* as

$$S = \iint_W |\mathbf{r}_u \times \mathbf{r}_v| \, du \, dv$$

where the integration is over the region W of the uv-plane which defines the whole of S. $d\sigma = |\mathbf{r}_u \times \mathbf{r}_v| \, du \, dv$ is said to be an *element of surface area.*

PROBLEM 6

Evaluate the surface area of the (curved part of the) circular cylinder

$$\mathbf{r} = (a \cos \theta, a \sin \theta, z) \qquad \theta \in [0, 2\pi], z \in [0, b]$$

From Problem 5 we have

$$|\mathbf{r}_\theta \times \mathbf{r}_z| = a$$

and so the total area of the curved part of the cylinder is

$$\int_0^b \int_0^{2\pi} a \, d\theta \, dz = 2\pi ab$$

as we would expect.

We can now define a *surface integral*. When we considered double integrals we essentially integrated a function over a plane surface (the xy-plane, for example). Now we are going to extend this to integration over a curved surface. Thus, let S be a simple surface, $\mathbf{r} = \mathbf{r}(u, v)$, and let W be the region in the uv-plane consisting of points corresponding to the points of S. Now let ϕ and $\mathbf{F}$ be scalar and vector fields defined on S, so that we can write $\phi = \phi(u, v)$ and $\mathbf{F} = \mathbf{F}(u, v)$. Then we define the surface integrals of ϕ and $\mathbf{F}$ over S to be

$$\iint_S \phi \, d\sigma = \iint_W \phi(u, v) \, | \, \mathbf{r}_u \times \mathbf{r}_v \, | \, du \, dv$$

$$\int_S \mathbf{F} \cdot d\sigma = \int_S \mathbf{F} \cdot \mathbf{n} \, d\sigma = \iint_W \mathbf{F}(u, v) \cdot (\mathbf{r}_u \times \mathbf{r}_v) \, du \, dv$$

Here, $d\sigma$ is interpreted as an element of surface area and $d\sigma$ is an abbreviation for $\mathbf{n} \, d\sigma$, where $\mathbf{n}$ is the unit normal vector to the surface S.

PROBLEM 7

Integrate $\phi(x, y, z) = x^2$ over the cone $z = \sqrt{x^2 + y^2}$, $z \in [0, 1]$.

The first step is to find a suitable parametrization of the cone. We can use cylindrical coordinates for this (*see* Fig. 10.4). Here, a typical point on the cone is $x = r \cos \theta$, $y = r \sin \theta$ and $z = \sqrt{x^2 + y^2} = r$, with $0 \le r \le 1$ and $0 \le \theta \le 2\pi$. This gives the parametrization

$$\mathbf{r}(r, \theta) = (r \cos \theta)\mathbf{i} + (r \sin \theta)\mathbf{j} + r\mathbf{k} \qquad r \in [0, 1], \theta \in [0, 2\pi]$$

From this we get

$$\mathbf{r}_r \times \mathbf{r}_\theta = -(r \cos \theta)\mathbf{i} - (r \sin \theta)\mathbf{j} + r\mathbf{k}$$

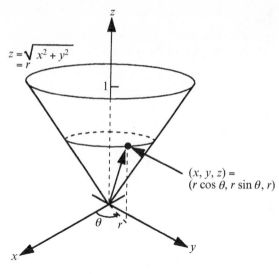

$$z = \sqrt{x^2 + y^2}$$
$$= r$$

$$(x, y, z) = (r\cos\theta, r\sin\theta, r)$$

Fig. 10.4

So

$$|\mathbf{r}_r \times \mathbf{r}_\theta| = \sqrt{2}\, r$$

Now

$$\iint_S \phi\, d\sigma = \int_0^{2\pi} \int_0^r \phi\, |\mathbf{r}_r \times \mathbf{r}_\theta|\, dr\, d\theta = \int_0^{2\pi} \int_0^1 x^2 \sqrt{2}\, r\, dr\, d\theta$$

$$= \int_0^{2\pi} \int_0^1 \sqrt{2}\, r^3 \cos^2\theta\, dr\, d\theta = \frac{\pi\sqrt{2}}{4}$$

It is possible that in working through Problem 7 your greatest difficulty arose in determining the parametrization. And you may also have wondered why we did not use the explicit form $z = \sqrt{x^2 + y^2}$ directly – as we did in the evaluation of surface area in Section 7.7, for example. We can indeed do this, but first, let us consider the form of the surface integral when we use an implicit representation, $f(x, y, z) = c$, of a surface. The way we approach this is to project a typical element of area onto an appropriate coordinate plane, so that we can express the surface integral as a double integral in two of the variables x, y, z.

Thus, let $\Delta\sigma_k$ be a patch of the surface which projects down onto a region ΔA_k of the coordinate plane – we can think of ΔA_k as the 'shadow' of $\Delta\sigma_k$ (*see* Fig. 10.5). We approximate the surface element $\Delta\sigma_k$ by a portion of the tangent plane ΔP_k covering the same part of the surface. To express the area of ΔP_k, and hence $\Delta\sigma_k$, in terms of the area of the shadow ΔA_k, we need the angle between ΔA_k and ΔP_k. This can be expressed in terms of the angle between the *normals* to these elementary plates. A unit normal vector, $\mathbf{p}$, to a portion, ΔA_k, of a coordinate plane is easy to find (e.g. for a portion of the xy-plane choose $\mathbf{p} = \mathbf{k}$). On the other hand, a ready-

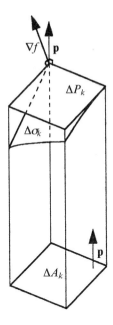

Fig. 10.5

made vector normal to the surface, and therefore to ΔP_k, is the gradient, ∇f. Now

$$|\nabla f . \mathbf{p}| = |\nabla f||\mathbf{p}||\cos \gamma|$$

where γ is the angle between the normals ∇f, $\mathbf{p}$ and therefore between the plates ΔP_k and ΔA_k. So we can write

$$\frac{1}{|\cos \gamma|} = \frac{|\nabla f|}{|\nabla f . \mathbf{p}|}$$

Now the relation between the area of ΔP_k and its projection ΔA_k is

$$\Delta P_k = \frac{\Delta A_k}{|\cos \gamma_k|}$$

where γ_k is the angle between the plates. Thus an approximation to the total surface area S is given by the sum

$$\sum \Delta P_k = \sum \frac{\Delta A_k}{|\cos \gamma_k|}$$

and in the limit we can write

$$S = \iint_R \frac{dA}{|\cos \gamma|}$$

where the integration is over the region R of the coordinate plane which is the 'shadow' of S. Using the expression obtained above for $1/|\cos \gamma|$, we can finally write the surface area as

$$S = \iint_R \frac{|\nabla f|}{|\nabla f \cdot \mathbf{p}|} \, dA$$

This is a very useful means of evaluating the surface area of a portion of the surface $f(x, y, z) = c$. It expresses the area as an integral over a coordinate plane and requires only the normal to the plane, and the implicit equation of the surface. The only caution required is in ensuring that $\cos \gamma$, or $\nabla f \cdot \mathbf{p}$, is non-zero. This may involve careful selection of the coordinate plane onto which we project. For example, suppose we are looking at a portion of the surface of the cylinder $f(x, y, z) = x^2 + y^2 = 1$. All surface elements in this case are perpendicular to the xy-plane, and $\cos \gamma = \nabla f \cdot \mathbf{k} \equiv 0$. So here we would not project onto the xy-plane, but possibly the xz- or yz-plane.

In general, then, we may have the following alternative forms for the surface area:

$$\iint_{R_{xy}} \frac{|\nabla f|}{|\nabla f \cdot \mathbf{p}|} \, dx \, dy \qquad \iint_{R_{xz}} \frac{|\nabla f|}{|\nabla f \cdot \mathbf{p}|} \, dx \, dz \qquad \iint_{R_{yz}} \frac{|\nabla f|}{|\nabla f \cdot \mathbf{p}|} \, dy \, dz$$

The obvious extension of the definition of the *surface integral* of a scalar field $\phi(x, y, z)$ over the surface S defined by $f(x, y, z) = c$ is

$$\iint_R \phi(x, y, z) \frac{|\nabla f|}{|\nabla f \cdot \mathbf{p}|} \, dA$$

The surface integral of a vector field, $\mathbf{F}$, or the *flux* of the field across an oriented surface S in the direction of the unit normal field $\mathbf{n}$ to the surface is

$$\iint_S \mathbf{F} \cdot \mathbf{n} \, d\sigma$$

Where S is defined implicitly by $f(x, y, z) = c$, we can take $\mathbf{n}$ to be one of the fields

$$\mathbf{n} = \pm \frac{\nabla f}{|\nabla f|}$$

depending on the chosen orientation. Then the flux may be written

$$\iint_S \mathbf{F} \cdot \mathbf{n} \, d\sigma = \iint_R \left(\mathbf{F} \cdot \frac{\pm \nabla f}{|\nabla f|} \right) \frac{|\nabla f|}{|\nabla f \cdot \mathbf{p}|} \, dA$$

$$= \iint_R \mathbf{F} \cdot \frac{\pm \nabla f}{|\nabla f \cdot \mathbf{p}|} \, dA$$

Finally, let us look at the form that the surface area and surface integrals take when the surface S is defined explicitly by, for example, $z = f(x, y)$. We can redefine this implicitly by defining a function $F(x, y, z) = f(x, y) - z$. We can then apply the above definitions to this (replacing $f(x, y, z)$ by $F(x, y, z)$). Thus we have

$$\nabla F = f_x \mathbf{i} + f_y \mathbf{j} - \mathbf{k}$$

The obvious projection is onto the xy-plane, so that $\mathbf{p} = \mathbf{k}$ and we get

$$\frac{|\nabla F|}{|\nabla F \cdot \mathbf{p}|} = \sqrt{f_x^2 + f_y^2 + 1}$$

giving for the surface area

$$\iint_R \sqrt{f_x^2 + f_y^2 + 1} \, dx \, dy$$

as found in Section 7.8. Similarly, for the integral of a scalar field $\phi(x, y, z)$ we have

$$\iint_R \phi(x, y, z) \sqrt{f_x^2 + f_y^2 + 1} \, dx \, dy$$

We can summarize our results by listing the surface elements appropriate to the various forms of representation of a surface:

- Explicit, $z = f(x, y)$:

$$d\sigma = \sqrt{f_x^2 + f_y^2 + 1} \, dx \, dy$$

- Implicit, $f(x, y, z) = c$

$$d\sigma = \frac{|\nabla f|}{|\nabla f \cdot \mathbf{p}|} \, dA$$

- Parametric, $\mathbf{r} = \mathbf{r}(u, v)$

$$d\sigma = |\mathbf{r}_u \times \mathbf{r}_v| \, du \, dv$$

PROBLEM 8

Repeat Problem 7 using the explicit representation of the surface.

The explicit form is $z = f(x, y) = \sqrt{x^2 + y^2}$, so

$$f_x = \frac{x}{\sqrt{x^2 + y^2}} \qquad f_y = \frac{y}{\sqrt{x^2 + y^2}}$$

and

$$\sqrt{f_x^2 + f_y^2 + 1} = \sqrt{2}$$

The required integral is therefore

$$\int_{-1}^{1} \int_{-\sqrt{1-y^2}}^{\sqrt{1-y^2}} x^2 \sqrt{2} \, dx \, dy = \frac{2\sqrt{2}}{3} \int_{-1}^{1} (1 - y^2)^{3/2} \, dy = \frac{2\sqrt{2}}{3} \cdot \frac{3\pi}{8} = \frac{\pi\sqrt{2}}{4}$$

as before.

1. Show that the unit normal to the sphere

$$\mathbf{r} = (a \cos \theta \sin \phi, \, a \sin \theta \sin \phi, \, a \cos \phi)$$

is

$$\mathbf{n} = (\cos \theta \sin \phi, \, \sin \theta \sin \phi, \, \cos \phi)$$

and show that the area of surface of the sphere cut off by $\theta = \theta_0$ and $\theta = \theta_0 + \alpha$ is $2\alpha a^2$.

2. Integrate $\phi(x, y, z) = x^2$ over the unit sphere $x^2 + y^2 + z^2 = 1$ by three different methods.

3. Calculate the flux of the vector field $\mathbf{F} = \mathbf{r}$ across the paraboloid $z = 1 - x^2 - y^2$, $z \geq 0$, in the outward direction by three different methods.

10.5 Gauss's divergence theorem

The divergence theorem is one of the central theorems of vector calculus. It relates the integral of a vector field $\mathbf{F}$ over a closed surface S to the volume integral of div $\mathbf{F}$ over the volume bounded by S. Physically this gives a relation between the flux of some quantity out of a bounded region and the amount of 'source material' inside the region. An obvious example is the relation between a volume of electric charge and the flux of the electric field it produces.

Thus, let τ be a closed region bounded by a simple-closed piecewise smooth surface S with outward normal $\mathbf{n}$. If the vector field $\mathbf{F}$ and its divergence are defined throughout τ, then the divergence theorem states that

$$\int_S \mathbf{F} . \, d\sigma = \iint_S \mathbf{F} . \, d\sigma = \iint_S \mathbf{F} . \, \mathbf{n} \, d\sigma = \iiint_\tau \nabla . \, \mathbf{F} \, d\tau \equiv \int_\tau \nabla . \, \mathbf{F} \, d\tau$$

(Note that in vector calculus it is common practice to use just a single integral sign for double and triple integrals.)

PROOF

For simplicity we will assume any line parallel to a coordinate axis (we take rectangular coordinates x, y, z) intersects S at most twice. As you are probably now well aware, there are means available for extending the proof to more general surfaces. Let the direction cosines of the outward unit normal be $\cos \alpha_1$, $\cos \alpha_2$, $\cos \alpha_3$, so that

$$\mathbf{n} = (\cos \alpha_1)\mathbf{i} + (\cos \alpha_2)\mathbf{j} + (\cos \alpha_3)\mathbf{k}$$

Then for $\mathbf{F} = F_1\mathbf{i} + F_2\mathbf{j} + F_3\mathbf{k}$ we have

$$\mathbf{F} . \mathbf{n} = F_1 \cos \alpha_1 + F_2 \cos \alpha_2 + F_3 \cos \alpha_3$$

and we proceed by proving that

$$\iint_S F_1 \cos \alpha_1 \, d\sigma = \iiint_\tau \frac{\partial F_1}{\partial x} \, dx \, dy \, dz$$

$$\iint_S F_2 \cos \alpha_2 \, d\sigma = \iiint_\tau \frac{\partial F_2}{\partial y} \, dx \, dy \, dz$$

$$\iint_S F_3 \cos \alpha_3 \, d\sigma = \iiint_\tau \frac{\partial F_3}{\partial z} \, dx \, dy \, dz$$

We illustrate the proof of the third equation, the others follow similarly.

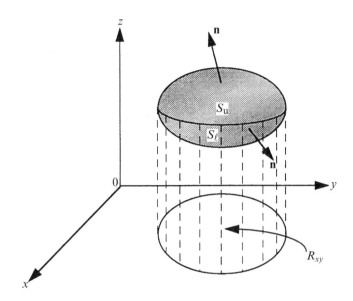

Fig. 10.6

Let R_{xy} denote the projection of τ onto the xy-plane. For any $(x, y) \in R_{xy}$ the line through (x, y) parallel to the z-axis cuts S in at most two points, an upper, P_u, and a lower, P_ℓ. As (x, y) varies over R_{xy}, P_u varies over the surface $z = f_u(x, y)$, and P_ℓ varies over the surface $z = f_\ell(x, y)$. Both f_u and f_ℓ are differentiable, and $S = S_\ell \cup S_u$. The solid τ is the set of all points with (*see* Fig. 10.6)

$$f_\ell(x, y) \le z \le f_u(x, y)$$

Let γ be the angle between the outward unit normal $\mathbf{n}$ and the positive z-axis. On S_u we have $\gamma = \alpha_3$, and so $\cos \alpha_3 \sec \gamma = 1$. On S_ℓ we have $\gamma = \pi - \alpha_3$, and so $\cos \alpha_3 \sec \gamma = -1$.

Now consider the element of surface area $d\sigma$ above the rectangle $dx \, dy$ in the xy-plane. The element is approximated by a parallelogram section of the tangent plane to the surface, with area given by

$$d\sigma = \frac{dx \, dy}{|\cos \gamma|}$$

from Section 10.4. So

$$\iint_{S_\ell} F_3 \cos \alpha_3 \, d\sigma = \iint_{R_{xy}} F_3 \cos \alpha_3 \sec \gamma \, dx \, dy = \iint_{R_{xy}} F_3(x, y, f_\ell(x, y)) \, dx \, dy$$

and

$$\iint_{S_u} F_3 \cos \alpha_3 \, d\sigma = \iint_{R_{xy}} F_3 \cos \alpha_3 \sec \gamma \, dx \, dy = \iint_{R_{xy}} F_3(x, y, f_u(x, y)) \, dx \, dy$$

Hence

$$\iint_S F_3 \cos \alpha_3 \, d\sigma = \iint_{S_u} F_3 \cos \alpha_3 \, d\sigma + \iint_{S_\ell} F_3 \cos \alpha_3 \, d\sigma$$

$$= \iint_{R_{xy}} (F_3(x, y, f_u(x, y)) - F_3(x, y, f_\ell(x, y))) \, dx \, dy$$

$$= \iint_{R_{xy}} \left(\int_{f_\ell(x, y)}^{f_u(x, y)} \frac{\partial F_3(x, y, z)}{\partial z} \, dz \right) dx \, dy$$

and by reversing the fundamental theorem of integral calculus

$$= \iiint_\tau \frac{\partial F_3(x, y, z)}{\partial z} \, dx \, dy \, dz$$

I will leave it for you as an exercise to confirm the other two equations. Adding the results produces the divergence theorem.

PROBLEM 9

> Let S be the surface of the cube with corners $(0, 0, 0)$, $(0, 0, 1)$, $(0, 1, 0)$, $(1, 0, 0)$, $(1, 1, 0)$, $(1, 0, 1)$, $(0, 1, 1)$, $(1, 1, 1)$, directed outwards. Use the divergence theorem to find the flux of the vector field $\mathbf{F} = 6x\mathbf{i} + 3y^2\mathbf{j} + z\mathbf{k}$ across S.

In this case $\nabla . \mathbf{F} = 7 + 6y$. So if τ is the solid cube enclosed by S, then the flux of $\mathbf{F}$ across S is

$$\iint_S \mathbf{F} . \mathbf{n} \, d\sigma = \iiint_\tau (7 + 6y) \, d\tau = \int_0^1 \int_0^1 \int_0^1 (7 + 6y) \, dz \, dy \, dx = 10$$

EXERCISES ON 10.5

1. Verify the first two integral relations in the proof of the divergence theorem.
2. Verify the divergence theorem for the field $\mathbf{F} = x\mathbf{i} + y\mathbf{j} + z\mathbf{k}$ over the sphere $x^2 + y^2 + z^2 = 1$.
3. If the scalar field ϕ together with its gradient are defined throughout a closed region τ bounded by a simple-closed surface S, then show that

$$\int_S \phi \, d\sigma = \int_\tau \nabla \phi \, d\tau$$

(*Hint:* Consider a vector field $\phi \mathbf{a}$, where $\mathbf{a}$ is an arbitrary constant vector.)

10.6 Stokes's theorem

Stokes's theorem relates the integral of a vector field **F** around a smooth, simple-closed curve C to the integral of curl **F** over any open oriented surface, S, bounded by C. Specifically, Stokes's theorem states that

$$\oint_C \mathbf{F} \cdot d\mathbf{r} = \int_S \text{curl } \mathbf{F} \cdot d\sigma$$

PROOF

I will only give a brief outline of a proof, leaving the details as a project. We consider the surface S as made up of many flat plates. There are two types of plate – those that are surrounded on all sides by others, and those that have one or more edges not adjacent to other plates. The boundary of S consists of the edges of the second kind of plates. A typical situation is illustrated in Fig. 10.7, in which the

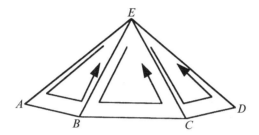

Fig. 10.7

triangles ABE, BCE, CDE represent a part of the surface S, with $ABCD$ forming part of the boundary of C. If we apply the circulation–curl form of Green's theorem to each of the triangles in turn and add the results, we get, putting it in vector language,

$$\left(\oint_{EAB} + \oint_{BCE} + \oint_{CDE} \right) \mathbf{F} \cdot d\mathbf{r} = \left(\iint_{EAB} + \iint_{BCE} + \iint_{CDE} \right) \nabla \times \mathbf{F} \cdot \mathbf{n} \, d\sigma$$

On the left-hand side the three integrals combine into a single integral around $ABCDE$, since integrals along the interior edges are taken twice each in opposite directions, thereby cancelling out. This is a standard trick in line and contour integration. It is the device by which what goes on *on* the boundary of a region is expressed in terms of what is happening *inside* the region.

So we have

$$\oint_{ABCDE} \mathbf{F} \cdot d\mathbf{r} = \iint_{ABCDE} \nabla \times \mathbf{F} \cdot \mathbf{n} \, d\sigma$$

Applying Green's theorem to all such plates and summing the results gives Stokes's theorem

$$\oint_C \mathbf{F} \cdot d\mathbf{r} = \int_S \text{curl } \mathbf{F} \cdot \mathbf{n} \, d\sigma$$

By taking the plates as small as we please we can extend this to a proof for any curved surface.

For practical purposes it is worth noting carefully the 'any . . . surface' in the statement of Stokes's theorem. Thus, a closed curve may form a boundary of any number of three-dimensional surfaces. The implication of Stokes's theorem is that the surface integral will have the same value for all such surfaces. However, there will naturally be surfaces for which the integration is most convenient, and you should always be on the look out to capitalize on this fact. Problem 10 illustrates the general idea.

Now let us return to our discussion of *conservative* or *gradient fields* in Section 10.2. For such a field we have $\mathbf{F} = -\nabla U$, so

$$\nabla \times \mathbf{F} = -\nabla \times \nabla U \equiv 0$$

modulo good behaviour of the potential function U. This provides a ready test for a conservative field – its curl must vanish. This being the case, again modulo good behaviour of $\mathbf{F}$ and the curve C, we can say from Stokes's theorem that if $\nabla \times \mathbf{F} = 0$ at every point of a simply-connected open region R, then on any piecewise smooth closed path C in R,

$$\oint_C \mathbf{F} . \, d\mathbf{r} = 0$$

This is another expression of the path independence of the work done by a conservative force, and forms an obvious link with our discussion in Section 10.2.

PROBLEM 10

Verify Stokes's theorem for the vector field $\mathbf{F} = z\mathbf{i} - 3x\mathbf{j} + 2y\mathbf{k}$, the surface S: $z = 4 - x^2 - y^2$, $z \geq 0$, and the boundary circle C: $x^2 + y^2 = 4$ (*see* Fig. 10.8). Assume that S is oriented in the positive z-direction.

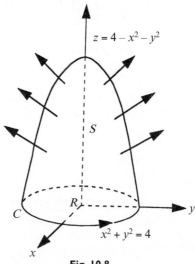

Fig. 10.8

From the orientation of S, the positive orientation of C is anti-clockwise and so it may be parametrized by

$$x = 2 \cos t \quad y = 2 \sin t \quad z = 0 \quad t \in [0, 2\pi]$$

Then $\mathbf{r}' = -2(\sin t)\mathbf{i} + 2(\cos t)\mathbf{j}$ and

$$\oint_C \mathbf{F} . \, d\mathbf{r} = -\int_0^{2\pi} 12 \cos^2 t \, dt = -12\pi.$$

Now consider the surface integral:

$$\iint_S (\nabla \times \mathbf{F}) . \, \mathbf{n} \, d\sigma = \iint_R (\nabla \times \mathbf{F}) . \, (2x\mathbf{i} + 2y\mathbf{j} + \mathbf{k}) \, dA$$

(note the choice of normal is correlated with the upward orientation of the surface)

$$= \iint_R (2\mathbf{i} + \mathbf{j} - 3\mathbf{k}) . \, (2x\mathbf{i} + 2y\mathbf{j} + \mathbf{k}) \, dA$$

$$= \iint_R (4x + 2y - 3) \, dA$$

$$= \int_0^{2\pi} \int_0^2 (4r \cos\theta + 2r \sin\theta - 3) \, r \, dr \, d\theta = -12\pi$$

as required by Stokes's theorem.   •

Clearly, in this case the line integral is much easier to integrate, and if viewed as a problem in finding the surface integral, then Stokes's theorem offers the opportunity to convert to the much easier line integral. On the other hand, if the problem is to find the line integral by using Stokes's theorem to convert it to a surface integral then much the easiest choice is to use the surface of the disc in the xy-plane. In this case $\mathbf{n} = \mathbf{k}$ and the surface integral becomes

$$\iint_R (-3) \, dA = \int_0^{2\pi} \int_0^2 (-3) \, r \, dr \, d\theta$$

$$= -12\pi$$

simplifying the calculations considerably.

EXERCISES ON 10.6

1. Find the circulation of the field $\mathbf{F} = (x^2 - y)\mathbf{i} + 4z\mathbf{j} + x^2\mathbf{k}$ around the curve C in which the plane $z = 2$ meets the cone $z = \sqrt{x^2 + y^2}$, anti-clockwise as viewed from above.
2. A simple-open surface S is enclosed within a correspondingly oriented simple-

closed curve C. If $\mathbf{a}$ is a constant vector field, and $\mathbf{r}$ denotes the position vector relative to the origin, prove that

$$\oint_C (\mathbf{a} \times \mathbf{r}) \cdot d\mathbf{r} = 2 \int_S \mathbf{a} \cdot d\boldsymbol{\sigma}$$

Deduce that

$$\int_S d\boldsymbol{\sigma} = \frac{1}{2} \oint_C \mathbf{r} \times d\mathbf{r}$$

FURTHER EXERCISES

1. If $\mathbf{F} = (x + 4)\mathbf{i} + xy\mathbf{j} + (xz - y)\mathbf{k}$, evaluate $\int_C \mathbf{F} \cdot d\mathbf{r}$ along the following paths C:
   (i) $x = 2t^2$, $y = t$, $z = t^3$ from $t = 0$ to $t = 1$;
   (ii) the straight line joining $(0, 0, 0)$ to $(2, 1, 1)$;
   (iii) the straight lines from $(0, 0, 0)$ to $(0, 0, 1)$, then to $(0, 1, 1)$, and then to $(2, 1, 1)$.

2. For the given vector field $\mathbf{F}$ and curve C, evaluate $\int_C \mathbf{F} \cdot d\mathbf{r}$:
   (i) $\mathbf{F} = xy\mathbf{i} + y^2\mathbf{j} - xz\mathbf{k}$
    C: $\mathbf{r}(t) = t\mathbf{i} - 2t\mathbf{j} - (\ln t)\mathbf{k}$ $\qquad\qquad t \in [1, 3]$
   (ii) $\mathbf{F} = x^2\mathbf{i} - 2xyz\mathbf{j} + z^2\mathbf{k}$
    C: $\mathbf{r}(t) = t\mathbf{i} + 2t\mathbf{j} - 4t\mathbf{k}$ $\qquad\qquad t \in [0, 3]$
   (iii) $\mathbf{F} = yz\mathbf{i} + xz\mathbf{j} + xy\mathbf{k}$
    C: $\mathbf{r}(t) = t\mathbf{i} + t^2\mathbf{j} + t^3\mathbf{k}$ $\qquad\qquad t \in [0, 1]$
   (iv) $\mathbf{F} = (2xz + \sin y)\mathbf{i} + x(\cos y)\mathbf{j} + x^2\mathbf{k}$
    C: $\mathbf{r}(t) = (\cos t)\mathbf{i} + (\sin t)\mathbf{j} + t\mathbf{k}$ $\qquad t \in [0, 2\pi]$
   (v) $\mathbf{F} = xy\mathbf{i} + (x^2 + y^2)\mathbf{j}$
    C: the parabola $y = x^2$ from $(1, 1)$ to $(2, 4)$
   (vi) $\mathbf{F} = -y\mathbf{i} + x\mathbf{j}$
    C: $y^2 = 3x$ from $(3, 3)$ to $(0, 0)$

3. Find the work done in moving a particle in the force field $\mathbf{F} = x^2\mathbf{i} + xy\mathbf{j} + z^2\mathbf{k}$ along
   (i) the straight line from $(0, 0, 0)$ to $(2, 1, 2)$;
   (ii) the circular helix $x = \cos t$, $y = \sin t$, $z = t$ from $t = 0$ to $t = 2\pi$;
   (iii) the curve defined by $x^2 = 4y$, $3x^3 = 8z$ from $x = 0$ to $x = 2$.

4. If $\mathbf{F} = (y + 2x)\mathbf{i} + (3x - 2y)\mathbf{j}$, compute the circulation of $\mathbf{F}$ about a circle C in the xy-plane with centre at the origin and radius 3, if C is traversed in the positive direction.

5. In each of the following cases show that the integral $\int_C \mathbf{F} \cdot d\mathbf{r}$ is independent of the curve C joining two given points by obtaining a scalar function ϕ such that $\mathbf{F} = \nabla\phi$. Find the value of the integral between the points given.
   (i) $\mathbf{F} = y^2\mathbf{i} + (2xy - e^y)\mathbf{j}$
    $(1, 0)$ to $(0, 1)$

(ii) $\mathbf{F} = 2xy^2\mathbf{i} + 2y(x^2 + 1)\mathbf{j}$
    $(0, 0)$ to $(2, 4)$

(iii) $\mathbf{F} = y^2z^4\mathbf{i} + 2xyz^4\mathbf{j} + 4xy^2z^3\mathbf{k}$
    $(0, 0, 0)$ to $(3, 2, 1)$

(iv) $\mathbf{F} = (2xy + 2z^2)\mathbf{i} + (x^2 - 3z)\mathbf{j} + (4xz - 3y)\mathbf{k}$
    $(-1, -1, -1)$ to $(1, 1, 1)$

(v) $\mathbf{F} = (4xy - 3x^2z^2)\mathbf{i} + 2x^2\mathbf{j} - 2x^3z\mathbf{k}$
    $(1, 0, 1)$ to $(1, 1, 1)$

6. Prove that $\mathbf{F} = r^2\mathbf{r}$ is conservative and find the scalar potential.

7. Determine whether the force field $\mathbf{F} = 2xz\mathbf{i} + (x^2 + y)\mathbf{j} + (2z + x^2)\mathbf{k}$ is conservative or non-conservative.

8. Verify Green's theorem for the following cases:

   (i) $\oint_C y^2 \, dx + x^2 \, dy$, where C is the square with vertices $(0, 0)$, $(1, 0)$, $(1, 1)$, $(0, 1)$ traversed anti-clockwise.

   (ii) $\oint_C x^2y \, dx + (x + y) \, dy$, where C is the unit circle, centre the origin.

   (iii) $\oint_C xy \, dx + (x - y) \, dy$, where C is the rectangle with vertices $(0, 1)$, $(1, 1)$, $(1, 3)$, $(0, 3)$ traversed anti-clockwise.

   (iv) $\oint_C xy \, dx + x^2 \, dy$, where C is the triangle with vertices $(0, 0)$, $(0, 1)$, $(1, 1)$.

9. Evaluate using Green's theorem

   (i) $\oint_C (3x + 4y) \, dx + (2x - 3y) \, dy$, where C is the unit circle with centre at the origin, traversed anti-clockwise.

   (ii) $\oint_C x^2 \, dy$, where C is the rectangle with vertices $(0, 0)$, $(a, 0)$, (a, b), $(0, b)$.

   (iii) $\oint_C \ln(1 + y) \, dx - xy/(1 + y) \, dy$, where C is the triangle with vertices $(0, 0)$, $(2, 0)$, $(0, 4)$ taken anti-clockwise.

   (iv) $\oint_C -y^2 \, dx + x \, dy$, where C is the square with vertices $(-1, -1)$, $(1, -1)$, $(1, 1)$, $(-1, 1)$ taken anti-clockwise.

   (v) $\oint_C 8y \, dx + xy \, dy$, where C is $\frac{1}{2}x^2 + \frac{1}{4}y^2 = 1$ taken anti-clockwise.

   (vi) $\oint_C (2x - y^3) \, dx - xy \, dy$, where C is the boundary of the region enclosed by circles $x^2 + y^2 = 1$ and $x^2 + y^2 = 9$.

   (vii) $\int_C (x^2 + y) \, dx + xy^2 \, dy$, where C is the closed curve defined by $y^2 = x$ and $y = x$.

   (viii) $\int_C x^2y \, dx + x \, dy$, where C is the triangular path through $(0, 0)$, $(1, 0)$, $(1, 2)$.

10. Evaluate $\iint_S \mathbf{F} \cdot \mathbf{n} \, d\sigma$ for each of the following cases:

    (i) $\mathbf{F} = 2y\mathbf{i} + x\mathbf{j} + z\mathbf{k}$ and S is the surface of the plane $2x + y = 8$ in the first octant cut off by the plane $z = 3$.

    (ii) $\mathbf{F} = y^2\mathbf{i} - (x + 2z)\mathbf{j} + yz\mathbf{k}$ and S is the surface of the plane $x + 2y + 2z = 6$ in the first octant.

    (iii) $\mathbf{F} = x^3\mathbf{i} + y^3\mathbf{j} + z^3\mathbf{k}$ and S is the sphere of radius a, centre the origin.

    (iv) $\mathbf{F} = z\mathbf{i} + \mathbf{j} - xy\mathbf{k}$ and S is the unit hemisphere centre the origin, for $z \geq 0$.

(v) $\mathbf{F} = z\mathbf{i} + \mathbf{j} - xy\mathbf{k}$ and S is the surface cut from the cylinder $y^2 + z^2 = 1$, $z \geq 0$, by the planes $x = 0$, $x = 2$.

(vi) $\mathbf{F} = x\mathbf{i} - y\mathbf{j} + xy\mathbf{k}$ and S is $\{(x, y, z) : x = \sqrt{y^2 + z^2}, y^2 + z^2 \leq 1\}$.

(vii) $\mathbf{F} = x\mathbf{k}$ and S is the part of the paraboloid $z = x^2 + y^2$ below the plane $z = y$, with downward unit normals.

(viii) $\mathbf{F} = \mathbf{r}$ and S is the portion of the cylindrical surface of radius a with axis along the z-axis between $z = 0$ and $z = \ell$.

(ix) $\mathbf{F} = 2y\mathbf{i} - z\mathbf{j} + x^2\mathbf{k}$ and S is the surface of the parabolic cylinder $y^2 = 8x$ in the first octant bounded by the planes $y = 4$ and $z = 6$.

11. Evaluate the given surface integrals:

(i) $\iint_S z^2 \, d\sigma$, where S is the portion of the cone $z = \sqrt{x^2 + y^2}$ between the planes $z = 1$ and $z = 2$.

(ii) $\iint_S 3y^2 z \, d\sigma$, where S is the portion of the cylinder $x^2 + y^2 = 1$ between the planes $z = 0$, $z = 1$, and on the positive x side of the yz-plane.

(iii) $\iint_S \sqrt{x^2 + y^2} \, d\sigma$, where S is $\{(x, y, z) : z = xy, 0 \leq x^2 + y^2 \leq 1\}$.

(iv) $\iint_S x^2 \, d\sigma$ where S is the surface of the cone $z = \sqrt{x^2 + y^2}$, $0 \leq z \leq 1$.

(v) $\iint_S (x + y) \, d\sigma$, where S is the portion of the plane $x + 2y - 3z = 4$ within $0 \leq x \leq 1$, $1 \leq y \leq 2$.

(vi) $\iint_S (x + 2y + 3z) \, d\sigma$, where S is the portion of the plane $x + y = 1$ in the first octant between $z = 0$ and $z = 1$.

12. Evaluate $\int_C \dfrac{-y \, dx + x \, dy}{x^2 + y^2}$ along the following paths:

(i) straight-line segments from $(1, 0)$ to $(1, 1)$, then to $(-1, 1)$, then to $(-1, 0)$;

(ii) straight line segments from $(1, 0)$, to $(1, -1)$, then to $(-1, -1)$, then to $(-1, 0)$. Show that although $\partial P/\partial y = \partial Q/\partial x$, the line integral *is dependent* on the path joining $(1, 0)$ to $(-1, 0)$ and explain.

13. Verify the divergence theorem for

(i) $\mathbf{F} = 2xz\mathbf{i} + yz\mathbf{j} + z^2\mathbf{k}$ over the upper half of the sphere $x^2 + y^2 + z^2 = a^2$.

(ii) $\mathbf{F} = x\mathbf{i} + y\mathbf{j} + z\mathbf{k}$ over the cylinder $x^2 + y^2 = 4$, $0 \leq z \leq 2$, including its top and base.

(iii) $\mathbf{F} = x^2\mathbf{i} - xz\mathbf{j} + z^2\mathbf{k}$ over the unit cube $0 \leq x \leq 1$, $0 \leq y \leq 1$, $0 \leq z \leq 1$.

(iv) $\mathbf{F} = 2x^2 y\mathbf{i} - y^2\mathbf{j} + 4xz^2\mathbf{k}$ over the region in the first octant bounded by $y^2 + z^2 = 9$ and $x = 2$.

14. Use the divergence theorem to evaluate $\iint_S \mathbf{F} \cdot \mathbf{n} \, d\sigma$ in the following cases:

(i) $\mathbf{F} = (y - x)\mathbf{i} + (z - y)\mathbf{j} + (y - x)\mathbf{k}$
S the cube bounded by the planes $x = \pm 1$, $y = \pm 1$, and $z = \pm 1$.

(ii) $\mathbf{F} = \mathbf{r}$
S the sphere of radius 2, centre the origin.

(iii) $\mathbf{F} = \mathbf{r}$
S the surface bounded by the paraboloid $z = 4 - x^2 - y^2$ and the xy-plane.

(iv) $\mathbf{F} = (x - z)\mathbf{i} + (y - x)\mathbf{j} + (z - y)\mathbf{k}$
 S the surface of the cylindrical solid bounded by $x^2 + y^2 = a^2$, $z = 0$, $z = 1$.
(v) $\mathbf{F} = (y^2 + z^2)^{3/2}\mathbf{i} + \sin[(x^2 - z^5)^{4/3}]\mathbf{j} + e^{x^2 - y^2}\mathbf{k}$
 S the ellipsoid $x^2/a^2 + y^2/b^2 + z^2/c^2 = 1$.
(vi) $\mathbf{F} = x^2\mathbf{i} + y^2\mathbf{j} + z^2\mathbf{k}$
 S the cylinder $0 \le x^2 + y^2 \le 4$, $0 \le z \le 4$, including the top and base.

15. If S is any closed surface enclosing a volume V and $\mathbf{F} = ax\mathbf{i} + by\mathbf{j} + cz\mathbf{k}$, prove
 that $\iint_S \mathbf{F} \cdot \mathbf{n}\, d\sigma = (a + b + c)V$.

16. Verify Stokes' theorem in the following cases:
 (i) $\mathbf{F} = (y - z + 2)\mathbf{i} + (yz + 4)\mathbf{j} - xz\mathbf{k}$, where S is the surface of the cube $x = 0$,
 $y = 0$, $z = 0$, $x = 2$, $y = 2$, $z = 2$, above the xy-plane.
 (ii) $\mathbf{F} = 3y\mathbf{i} - 2y\mathbf{j} + z^2\mathbf{k}$
 S the hemisphere $x^2 + y^2 + z^2 = 1$, $z \ge 0$.
 (iii) $\mathbf{F} = 2z\mathbf{i} + 3x\mathbf{j} + 5y\mathbf{k}$
 S is the paraboloid $z = 4 - x^2 - y^2$ above the xy-plane, and C is the circle
 $x^2 + y^2 = 4$.
 (iv) $\mathbf{F} = -3y\mathbf{i} + 3x\mathbf{j} + z^4\mathbf{k}$
 S is the portion of the ellipsoid $2x^2 + 2y^2 + z^2 = 1$ that lies above the plane
 $z = 1/\sqrt{2}$.
 (v) $\mathbf{F} = xz\mathbf{i} - y\mathbf{j} + x^2y\mathbf{k}$
 S is the surface of the region bounded by $x = 0$, $y = 0$, $z = 0$, $2x + y + 2z = 8$
 which is not included in the xz-plane.

17. Use Stokes's theorem to evaluate $\oint_C \mathbf{F} \cdot \mathbf{dr}$ for the following:
 (i) $\mathbf{F} = y\mathbf{i} + xz\mathbf{j} + x^2\mathbf{k}$
 C the boundary of the portion of the plane $x + y + z = 1$ in the first octant,
 traversed anti-clockwise when viewed from above.
 (ii) $\mathbf{F} = (z - 2y)\mathbf{i} + (3x - 4y)\mathbf{j} + (z + 3y)\mathbf{k}$
 C the unit circle in the plane $z = 2$.
 (iii) $\mathbf{F} = xy\mathbf{i} + x^2\mathbf{j} + z^2\mathbf{k}$
 C the intersection of the paraboloid $z = x^2 + y^2$ and the plane $z = y$,
 traversed anti-clockwise looking down the positive z-axis.
 (iv) $\mathbf{F} = (2x - y)\mathbf{i} - yz^2\mathbf{j} - y^2z\mathbf{k}$
 C the circle $x^2 + y^2 = 1$, $z = 0$.

18. Evaluate $\iint_S (\nabla \times \mathbf{F}) \cdot \mathbf{n}\, d\sigma$, where $\mathbf{F} = (x^2 + y - 4)\mathbf{i} + 3xy\mathbf{j} + (2xz + z^2)\mathbf{k}$, and S is
 the surface of (i) the hemisphere $z^2 + y^2 + z^2 = 16$ above the xy-plane, (ii) the
 paraboloid $z = 4 - (x^2 + y^2)$ above the xy-plane.

19. S is a simple-open surface, bounded by a correspondingly oriented curve C. If
 ϕ, ψ are continuously differentiable scalar fields, use Stokes's theorem to show
 that

$$\oint_C \phi \nabla \psi \cdot \mathbf{dr} = \int_\tau \nabla \phi \times \nabla \psi \cdot d\sigma$$

20. Write Stokes's theorem in terms of rectangular coordinates.

21. The scalar field ϕ has continuous second-order derivatives in the closed region τ bounded by a simple-closed surface S. Prove that

$$\int_S \phi \nabla\phi \cdot d\boldsymbol{\sigma} = \int_\tau \{\phi\nabla^2\phi + (\nabla\phi)^2\}d\tau$$

If $\phi = x + y + z$, deduce that the left-hand side is equal to three times the volume enclosed within the surface S.

22. Essay and discussion topics:

(i)   The connection between the theorems of Green, Gauss and Stokes.
(ii)   The different, equivalent forms for the line integral.
(iii)   Gradient fields, the fundamental theorem of integral calculus, and the conservation of energy.
(iv)   The different forms of Green's theorem.
(v)   Three ways to do a surface integral.
(vi)   Applications of the divergence theorem.
(vii)   Stokes's theorem and conservative forces.
(viii)   Applications of Stokes's theorem.

Bibliography

There are a number of books on vector calculus, but few that cover the full range of multi-variable calculus from partial differentiation to the integral theorems. Limitations of space mean that I am also unable to do full justice to this range, and so I have sacrificed some depth and detail in places. However, there are many books that you can dip into to supplement these – particularly the hefty American texts of Anton, Salas and Hille, Grossman, Larson *et al.*, and Thomas and Finney. These giants contain many worked examples and you may find that consulting these pays dividends when working through the exercises I have set. Other books are more specialized, and will be able to expand on such things as proofs, rigour and applications, particularly of vector calculus.

Anton, H. (1995) *Calculus with analytic geometry* (5th edn), John Wiley, New York.

Anton, H. & Rorres, C. (1991) *Elementary linear algebra: applications version* (6th edn), John Wiley, New York.

Bourne, D.E. & Kendall, P.C. (1977) *Vector analysis and Cartesian tensors* (2nd edn), Thomas Nelson, Sunbury-on-Thames.

Cox, W. (1996) *Ordinary Differential Equations*, Arnold, London.

Dineen, S. (1995) *Functions of two variables*, Chapman & Hall, London.

Durrant, A. (1996) *Vectors in physics and engineering*, Chapman & Hall, London.

Grossman, S.I. (1988) *Calculus* (4th edn), Harcourt Brace Jovanovich, New York.

Hirst, A.E. (1995) *Vectors in 2 or 3 Dimensions*, Arnold, London.

Jones, D.S. and Jordan, D.W. (1970) *Introductory analysis, Volume 2*, John Wiley, London.

Kopp, P.E. (1996) *Analysis*, Arnold, London.

Larson, R.E., Hostetler, R.P. and Edwards, B.H. (1994) *Calculus with analytic geometry*, D.C. Heath, Toronto.

Marsden, J.E. and Tromba, A.J. (1996) *Vector calculus* (4th edn), W.H. Freeman, New York.

Pearson, D. (1996) *Calculus and ODEs*, Arnold, London.

Pedrick, G. (1994) *A first course in analysis*, Springer-Verlag, New York.

Salas, S.L. and Hille, E. (1995) *Calculus: one and several variables* (4th edn), John Wiley, New York.

Schey, H.M. (1996) *Div, Grad, Curl and All That* (3rd edn), W.W. Norton, New York.

Spiegel, M.R. (1974) *Vector analysis*, Shaum's Outline Series, McGraw-Hill, New York.

Stewart, I. (1996) *From here to infinity*, Oxford University Press, Oxford.

Thomas Jr, G.B. and Finney, R.L. (1996) *Calculus and Analytic Geometry*, Addison-Wesley, New York.

Solutions to Exercises

Chapter I

EXERCISES ON 1.1

1. $\partial f/\partial x = 2xy\cos(x+y) - yx^2\sin(x+y)$
   $\partial f/\partial y = x^2\cos(x+y) - yx^2\sin(x+y)$
2. 0.04

EXERCISES ON 1.2

1. $z = 3$
2. $f_{xx} = 2/(x+2y)^3, f_{yy} = 8/(x+2y)^2, f_{xy} = 4/(x+2y)^3, x+2y \neq 0$

EXERCISES ON 1.3

1. $$\int_a^b \int_c^d f(x, y)\, dy\, dx = \int_c^d \int_a^b f(x, y)\, dx\, dy$$

2. 11/2. This is the volume under the plane $z = x + 2y$ over the rectangle $0 \leq x \leq 1$, $2 \leq y \leq 3$.

EXERCISES ON 1.4

1. (i) $2xy\mathbf{i} + (x^2 + 2yz)\mathbf{j} + y^2\mathbf{k}$
   (ii) $yz\mathbf{i} + xz\mathbf{j} + xy\mathbf{k}$
2. (i) div $= x + y + z$ curl $= -y\mathbf{i} - z\mathbf{j} - x\mathbf{k}$
   (ii) div $= 3$ curl $= -\mathbf{i} - \mathbf{j} - \mathbf{k}$

FURTHER EXERCISES

2. $\partial z/\partial x = 2x$ $\partial z/\partial y = 2y$ (i) $2x$ (ii) $2y$
3. 0.4%
4. (i) $f_x = 2xy^2, f_y = 2z^2y, f_{xx} = 2y^2, f_{yy} = 2x^2, f_{xy} = 4xy = f_{yx}$
   (ii) $f_x = y\sin(xy), f_y = -x\sin(xy), f_{xx} = -y^2\cos(xy)$
        $f_{yy} = -x^2\cos(xy), f_{xy} = -\sin(xy) - xy\cos(xy) = f_{yx}$
5. $f_x(1, 2) = f_y(1, 2) = 0$. The tangent plane to the surface is horizontal: the point *may* be a maximum or a minimum, but not necessarily – it may be a *saddle point*.
6. 20/3
7. (i) $(3 + y)\mathbf{i} + (2z + x)\mathbf{j} + 2y\mathbf{k}$
   (ii) $(y\cos(xz) - xyz\sin(xz))\mathbf{i} + x\cos(xz)\mathbf{j} - x^2y\sin(xz)\mathbf{k}$
8. (i) div $= 2xy^2 + 2xze^{yz} + 1$
        curl $= -2xye^{yz}\mathbf{i} + (2e^{yz} - 2x^2y)\mathbf{k}$
   (ii) div $= e^y(e^x + e^z)$

Chapter 2

EXERCISES ON 2.1

1. (i) (a) 1   (b) $e^{-1}\cos 1$ (c) $e^{-1}\cos 1$
   (ii) (a) 2   (b) 3   (c) 5

2.

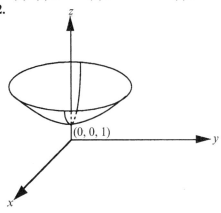

EXERCISES ON 2.2

1. (i) Closed $(0, 0)$, $(1, 1)$ (ii) Open $(2, 2)$, $(0, 1)$
   (iii) Neither $(1, 1)$, $(0, 1)$ (iv) Open $(\frac{1}{2}, 0, 0)$, $(0, 0, 2)$
   (v) Closed $(0, 0, 0)$, $(1, 1, 0)$ (vi) Open $(0, 0)$, $(1, 1)$
2. (i) Bounded and connected – closed region
   (ii) Unbounded and connected – open region
   (iii) Bounded and connected – region
   (iv) Unbounded and connected – region
   (v) Bounded and connected – region
   (vi) Unbounded, not connected – not a region

EXERCISE ON 2.4

(i)

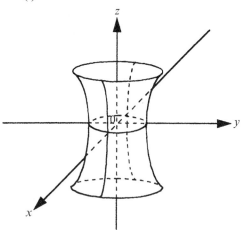

(ii)

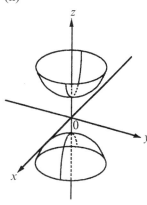

FURTHER EXERCISES

1. (i) $0, 1, -\pi, 0$ (ii) $0, 5/16, 1/2, 2$
    (iii) $0, 0, 4, -1/\sqrt{2}$

2. (i) $\mathbb{R}^2, [-1, 1]$
    (ii) $\{(x, y) : x + y + 2 \geq 0\}, [0, \infty)$
    (iii) $\{(x, y) : x \neq 0\}, \mathbb{R}$
    (iv) $\mathbb{R}^2, (0, \infty)$
    (v) $\mathbb{R}^2, [0, \infty)$
    (vi) $\mathbb{R}^2, \mathbb{R}$
    (vii) $\{(x, y) : y > x^2 - 1\}, \mathbb{R}$
    (viii) $\{(x, y) : x^2 \geq y^2 \text{ and } x \neq y\}, [0, \infty)$
    (ix) $\{(x, y, z) : x + 2y + 3z \geq 0\}, [0, \infty)$
    (x) $\{(x, y, z) : x + y \neq 0\}, \mathbb{R}$
    (xi) $\{(x, y, z) : y \neq 0\}, [-1, 1]$
    (xii) $\{(x, y, z) : x + y + 2z + 1 > 0\}, \mathbb{R}$

3. (i)

    (ii) Elliptic paraboloid
    (iii) Quadratic cone
    (iv) Hyperboloid of two sheets

4. (i) Straight lines of slope 1: $y = x - z$
    (ii) The y-axis and the lines $y = [(1 - z)/z]x$, with the origin omitted
    (iii) Parallel straight lines of slope -1, $y = -x + (z^2 - 1)$
    (iv) Concentric ellipses centred on the origin with equations $x^2 + 4y^2 = 1 - z^2$ if $z < 1$; if $z = 1$, the level 'curve' is simply the origin.

5. (i) $(x - 2)^2 + (y + 1)^2 < 4$
    (ii) $(x + 1)^2 + (y + 1)^2 \leq 16$
    (iii) $(x + 1)^2 + (y - 2)^2 + z^2 \leq 9$
    (iv) $(x - 1)^2 + (y - 1)^2 + z^2 \leq 1$

6. (i) Interior is the rectangle contained within the lines $x = 1$, $x = 3$, $y = 2$, $y = 4$. These lines form the boundary, and the set is closed.
    (ii) Interior is the rectangle inside $x = a$, $x = b$, $y = c$, $y = d$, which lines form the boundary. The set is open.
    (iii) The interior is the annulus between the circles centre the origin and with radii 1 and 3. The circles form the boundary. The set is open.
    (iv) As for (iii) but with circles of radii $\sqrt{2}$ and $\sqrt{3}$. The set is neither open nor closed.

(v) The interior is the elliptic annulus between the ellipses $x^2 + 2y^2 = 1$ and $x^2 + 2y^2 = 5$. These ellipses form the boundary of the set, which is open.

(vi) The interior is the entire region below the paraboloid $y = x^2$, which forms the boundary of the region, which is closed.

(vii) Interior is the inside of the box with the sides $x = 1$, $x = 2$, $y = 2$, $y = 3$, $z = 0$, $z = 1$, which planes form the boundary. The set is open.

(viii) As (vii) for the interior and boundary, but neither open nor closed.

(ix) Interior is the inside of the cylinder of radius 2, axis along the z-axis, between $z = 0$ and $z = 3$. The planes $z = 0$ and $z = 3$, with the portion of the cylinder between them, define the boundary. The set is closed.

(x) The interior is the inside of the sphere radius 2, centred on $(1, -2, 3)$. The sphere is the boundary of the set, which is open.

7. (i) $r^2 + z^2 = 4$ (ii) $r^2 = 1$
    (iii) $z = 3r$ (iv) $r^2 \cos^2 \theta + z^2 = 9$

8.

	Rectangular	Cylindrical	Spherical
(i)	$(1, 1, 1)$	$\left(\sqrt{2}, \frac{\pi}{4}, 1\right)$	$\left(\sqrt{3}, \frac{\pi}{4}, \frac{1}{\sqrt{3}}\right)$
(ii)	$(\sqrt{2}, \sqrt{6}, 2\sqrt{2})$	$\left(2\sqrt{2}, \frac{\pi}{3}, 2\sqrt{2}\right)$	$\left(4, \frac{\pi}{3}, \frac{\pi}{4}\right)$
(iii)	$(-1, \sqrt{3}, 6)$	$\left(2, \frac{2\pi}{3}, 6\right)$	$\left(2\sqrt{10}, \frac{2\pi}{3}, \cos^{-1}\left(\frac{3}{\sqrt{10}}\right)\right)$
(iv)	$(\sqrt{2}, -\sqrt{2}, 1)$	$\left(2, \frac{7\pi}{4}, 1\right)$	$\left(\sqrt{5}, \frac{7\pi}{4}, \tan^{-1}(2)\right)$
(v)	$(0, -2\sqrt{2}, 0)$	$\left(2\sqrt{2}, \frac{3\pi}{2}, 0\right)$	$\left(2\sqrt{2}, \frac{\pi}{2}, \frac{3\pi}{2}\right)$
(vi)	$(2, 2\sqrt{3}, -3)$	$\left(4, \frac{\pi}{3}, -3\right)$	$\left(5, \frac{\pi}{3}, \pi - \tan^{-1}\left(\frac{4}{3}\right)\right)$

9. $z = r^2 + 2r \cos \theta + r \sin \theta$

10. $\rho = 2 \cos \phi$

11. (i) $x^2 + (y - 2)^2 = 4$ (ii) $(x - 1)^2 + y^2 = 1$

12. (i) Circular cylinder, $x^2 + y^2 = 1$, radius 1 and axis along the z-axis.
    (ii) Horizontal plane one unit above the xy-plane.

13. (i) Quadric cone   (ii) Parabolic cylinder
    (iii) Hyperboloid of two sheets
    (iv) Sphere radius 2, centre the origin
    (v) Elliptic paraboloid   (vi) Hyperbolic paraboloid

14. (i) Elliptic cylinder centred on the x-axis
    (ii) Parabolic cylinder tangent to the x-axis
    (iii) Hyperbolic cylinder 'parallel' to the z-axis
    (iv) Elliptic paraboloid
    (v) Quadric cone
    (vi) Hyperboloid of two sheets
    (vii) Elliptic paraboloid

(viii) Paraboloid of revolution
(ix) Ellipsoid
(x) Hyperboloid of one sheet
(xi) Hyperboloid of two sheets
(xii) Elliptic paraboloid
(xiii) Hyperbolic paraboloid

Chapter 3

EXERCISES ON 3.1

2. (i) $9 + y^2$ (ii) $2y$

EXERCISES ON 3.2

1. 1

2. (i) 1 (ii) $\sqrt{\dfrac{2}{3}}$ (iii) 1

3. 0

4. (i) Does not exist (ii) Does not exist

EXERCISES ON 3.3

1. (i) Not continuous (ii) Continuous

2. Not continuous

EXERCISES ON 3.4

1. (i) $f_x = y, f_y = x$ (ii) $f_x = 1, f_y = 2$
(iii) $f_x = 6xy - 2y^2, f_y = 3x^2 - 4xy$
(iv) $f_x = -y\sin(xy), f_y = -x\sin(xy)$
(v) $f_x = ye^{xy}, f_y = xe^{xy}$

2. Symmetry in x and y provides a short-cut.

EXERCISES ON 3.5

1. $\dfrac{\partial f(x, g(y))}{\partial y} = \dfrac{\partial g}{\partial y} \dfrac{\partial f(x, g)}{\partial g}$

2. (i) $f_x = 2, f_y = 3$
(ii) $f_x = 2xy, f_y = x^2$
(iii) $f_x = \dfrac{y^2}{(x + y)^2}, f_y = \dfrac{x^2}{(x + y)^2}$
(iv) $f_x = 2xy^2e^{x^2}, f_y = 2ye^{x^2}$

FURTHER EXERCISES

1. (i) 1 (ii) 0 (iii) 2

2. (i) $x - 2, x \neq 2$ (ii) $\dfrac{x-1}{x^2+1}$ (iii) x

(iv) $\sqrt{x+1}, x \neq 1$ (v) $x - 2, x \neq 2$

3. (i) 4   (ii) 1   (iii) $2a + b$ (iv) 1/3   (v) 3

4. (i) 4/7   (ii) 4/7   (iii) 0   (iv) $\sin 1$

(v) 1   (vi) 0   (vii) 19   (viii) 1

6. (i) 0   (ii) 8/3   (iii) 0

8. (i) Continuous   (ii) Continuous   (iii) Not continuous

(iv) Continuous   (v) Continuous   (vi) Not continuous

9. Define $f(0, 0) = 0$

10. (i) Not continuous   (ii) Continuous   (iii) Not continuous

(iv) Continuous

11. (i) $\{(x, y) : x \leq y\}$ (ii) $\{(x, y) : x \neq y\}$ (iii) $\{(x, y) : 2x - 3y + 3 > 0\}$

(iv) $\mathbb{R}^2$ (v) $\{(x, y) : \dfrac{x^2}{4} + y^2 < 1\}$

(vi) $\{(x, y, z) : x + 2y - z + 2 \neq 0\}$

(vii) $\{(x, y, z) : -1 \leq z^2 + y \leq 1\}$

(viii) $\{(x, y, z) : x^2 + y^2 + z^2 < 1\}$

12. $c = 0$

13. (i) 2   (ii) 0

14. $g(x) = 2x$

15. (i) $f_x = 2, f_y = 1$ (ii) $f_x = 2x, f_y = 6y$

(iii) $f_x = -\dfrac{1}{x^2 y}, f_y = -\dfrac{1}{xy^2}$

(iv) $f_x = y \cos(xy), f_y = x \cos(xy)$ (v) $f_x = e^{x+y}, f_y = e^{x+y}$

16. (i) $f_x = f_y = 0$ at $(0, 0)$

Chapter 4

EXERCISE ON 4.1

(i) $f_x = \dfrac{y^2 \tan(x + y) + xy(x + y) \sec^2(x + y)}{(x + y)^2}$

$f_y = \dfrac{x^2 \tan(x + y) + xy(x + y) \sec^2(x + y)}{(x + y)^2}$

(ii) $f_x = (3 - y \sin(xy)) e^{3x + \cos(xy)}$

$f_y = -x \sin(xy) e^{3x + \cos(xy)}$

(iii) $f_x = \dfrac{2x^3 y^2 + 3x^2 - y^3}{(xy^2 + 1)^2}, f_y = \dfrac{xy^2 + 1 - 2yx^3 - 2y^2}{(xy^2 + 1)^2}$

EXERCISES ON 4.2

1. (i) $f_x = 3x^2 y^2 + 4y^4, f_y = 2x^3 y + 16xy^3$

$f_{xx} = 6xy^2, f_{yy} = 2x^3 + 48xy^2, f_{xy} = 6x^2 y + 16y^3$

(ii) $f_x = \dfrac{2x}{x^2 + y^2}, f_y = \dfrac{2y}{x^2 + y^2}$

$f_{xx} = \dfrac{2(y^2 - x^2)}{(x^2 + y^2)^2}, f_{yy} = \dfrac{2(x^2 - y^2)}{(x^2 + y^2)^2}, f_{xy} = \dfrac{-4xy}{(x^2 + y^2)^2}$

(iii) $f_x = e^{xy}(y \cos(x + y) - \sin(x + y))$,
$\quad f_y = e^{xy}(x \cos(x + y) - \sin(x + y))$
$\quad f_{xx} = e^{xy}(y^2 - 1) \cos(x + y), \ f_{yy} = e^{xy}(x^2 - 1)\cos(x + y)$
$\quad f_{xy} = xye^{xy} \cos(x + y) - (x + y)e^{xy} \sin(x + y)$

EXERCISES ON 4.3

1. $f_{xxx} = e^{x-y}(3 + x + \sin z), \qquad f_{xxy} = -e^{x-y}(2 + x + \sin z)$
$\quad f_{xxz} = e^{x-y} \cos z, \qquad f_{xyz} = -e^{x-y} \cos z$
$\quad f_{yyx} = e^{x-y}(1 - x - \sin z), \qquad f_{yyy} = -e^{x-y}(x + \sin z)$
$\quad f_{yyz} = e^{x-y}\cos z, \qquad f_{zzx} = -e^{x-y} \sin z$
$\quad f_{zzy} = e^{x-y}\sin z, \qquad f_{zzz} = -e^{x-y} \cos z$

2. nf

EXERCISE ON 4.4

(i) We obtain the ODEs
$$f''(x) - \lambda f(x) = 0, \ g''(y) - \lambda g(y) = 0$$
(ii) $f''(x) - \lambda f(x) = 0, \ g'(y) - \lambda g(y) = 0$

EXERCISES ON 4.5

1. $\dfrac{\partial f}{\partial r} = \dfrac{\partial f}{\partial x}\dfrac{\partial x}{\partial r} + \dfrac{\partial f}{\partial y}\dfrac{\partial y}{\partial r} + \dfrac{\partial f}{\partial z}\dfrac{\partial z}{\partial r}$

$\quad \dfrac{\partial f}{\partial s} = \dfrac{\partial f}{\partial x}\dfrac{\partial x}{\partial s} + \dfrac{\partial f}{\partial y}\dfrac{\partial y}{\partial s} + \dfrac{\partial f}{\partial z}\dfrac{\partial z}{\partial s}$

2. $e^{t+t^2}((1 + 2t) \cos(t^3) - 3t^2 \sin(t^3))$

3. $\dfrac{\partial z}{\partial r} = -2rs^2 \sin r \cos r \sin s + s^2 \sin s \cos^2 r$

$\quad \dfrac{\partial z}{\partial s} = 2rs \cos^2 r \sin s + rs^2 \cos^2 r \cos s$

4. $-\dfrac{3x^2 + 2y \sin(x + y) + 2xy \cos(x + y)}{2x \sin(x + y) + 2xy \cos(x + y)}$

EXERCISE ON 4.6

6.00333 to five decimal places.

FURTHER EXERCISES

1. (i) $f_x = 1, f_y = 3, f_{xx} = f_{yy} = f_{xy} = 0$
$\quad$ (ii) $f_x = 4x, f_y = 8y, f_{xx} = 4, f_{yy} = 8, f_{xy} = 0$
$\quad$ (iii) $f_x = 12x^3 - 8xy^2, f_y = -8x^2y + 15y^2$
$\quad\quad f_{xx} = 36x^2 - 8y^2, f_{yy} = -8x^2 + 30y, f_{xy} = -16xy$
$\quad$ (iv) $f_x = (y + xy)e^{x+y}, f_y = (x + xy)e^{x+y}$
$\quad\quad f_{xx} = (2y + xy)e^{x+y}, f_{yy} = (2x + xy)e^{x+y}$
$\quad\quad f_{xy} = (1 + x + y + xy)e^{x+y}$

(v) $f_x = -\sin(x + y^2), f_y = -2y \sin(x + y^2)$
$f_{xx} = -\cos(x + y^2), f_{yy} = -2 \sin(x + y^2) - 2y \cos(x + y^2)$
$f_{xy} = -2y \cos(x + y^2)$

(vi) $f_x = \cos x \cos(xy) - y \sin x \sin(xy)$
$f_y = -x \sin x \sin(xy)$
$f_{xx} = -\sin x \cos(xy) - 2y \cos x \sin(xy) - y^2 \sin x \sin(xy)$
$f_{yy} = -x^2 \sin x \cos(xy)$
$f_{xy} = -x \cos x \sin(xy) - \sin x \sin(xy) - xy \sin x \cos(xy)$

(vii) $f_x = \dfrac{2y}{(x + y)^2}, f_y = -\dfrac{2x}{(x + y)^2}$

$f_{xx} = -\dfrac{4y}{(x + y)^3}, f_{yy} = -\dfrac{4x}{(x + y)^3}, f_{xy} = \dfrac{2(x - y)}{(x + y)^3}$

(viii) $f_x = \dfrac{2x}{x^2 + y} - \dfrac{2}{2x + 3y}, f_y = \dfrac{1}{x^2 + y} - \dfrac{3}{2x + 3y}$

$f_{xx} = \dfrac{2(y - x^2)}{(x^2 + y)^2} + \dfrac{4}{(2x + 3y)^2}$

$f_{yy} = -\dfrac{1}{(x^2 + y)^2} + \dfrac{9}{(2x + 3y)^2}$

$f_{xy} = -\dfrac{2x}{(x^2 + y)^2} + \dfrac{6}{(2x + 3y)^2}$

(ix) $f_x = \dfrac{x + y - 1}{(x + y)^2} e^{x+y} = f_y$

$f_{xx} = \left[\dfrac{1}{x + y} - \dfrac{2}{(x + y)^2} + \dfrac{2}{(x + y)^3}\right] \dfrac{e^{x+y}}{x + y} = f_{yy} = f_{xy}$

(x) $f_x = e^{x+y}(\cos(x^2 + y) - 2x \sin(x^2 + y))$
$f_y = e^{x+y}(\cos(x^2 + y) - \sin(x^2 + y))$
$f_{xx} = e^{x+y}((1 - 4x^2) \cos(x^2 + y) - 2(1 + 2x) \sin(x^2 + y))$
$f_{yy} = -2e^{x+y} \sin(x^2 + y)$
$f_{xy} = e^{x+y}((1 - 2x) \cos(x^2 + y) - (1 + 2x) \sin(x^2 + y))$

2. (i) $f_{xx} = 0, f_{yy} = 2, f_{zz} = 6z, f_{xy} = f_{xz} = f_{yz} = 0$

(ii) $f_{xx} = -\dfrac{(y^2 z + yz^2)}{(x + y + z)^3}, f_{yy} = -\dfrac{(x^2 z + xz^2)}{(x + y + z)^3}$

$f_{zz} = -\dfrac{(x^2 y + xy^2)}{(x + y + z)^3},$

$f_{xy} = \dfrac{2xyz + z^2(x + y + z)}{(x + y + z)^3}$

f_{xz} – replace the z^2 by y^2
f_{yz} – replace the z^2 by x^2

(iii) $f_{xx} = e^{x+y^2+z^3}, f_{yy} = (4y^2 + 2)e^{x+y^2+z^3}$
$f_{zz} = (9z^4 + 6z) e^{x+y^2+z^3}$
$f_{xy} = 2ye^{x+y^2+z^3}, f_{xz} = 3z^2 e^{x+y^2+z^3}$
$f_{yz} = 6z^2 ye^{x+y^2+z^3}$

(iv) With $u = (x + y + 2z)^2, v = x^2 + y^2$

$f_{xx} = -\dfrac{1}{u} - \dfrac{2(x^2 - y^2)}{v}, f_{yy} = -\dfrac{1}{u} - \dfrac{2(y^2 - x^2)}{v}, f_{zz} = -\dfrac{4}{u}$

$f_{xy} = -\dfrac{1}{u} + \dfrac{4xy}{v^2}, f_{xz} = -\dfrac{2}{u} = f_{yz}$

(v) $f_{xx} = \dfrac{2z}{x^3}, f_{yy} = \dfrac{2x}{y^3}, f_{zz} = \dfrac{2y}{z^3}$

$f_{xy} = -\dfrac{1}{y^2}, f_{xz} = -\dfrac{1}{x^2}, f_{yz} = -\dfrac{1}{z^2}$

6. 0

7. $n = -3, 2$

11. (i) $2 \cos t - 8 \sin t + 2 \cos 2t$
   (ii) $14t^6 + 8t^3 - 12t^{11}$
   (iii) 5.3π cm s^{-1}

12. (i) Pressure decreases by 2%   (ii) (a) $2\varepsilon\%$ (b) $\varepsilon\%$
   (iii) 0.0305 units

17. $\dfrac{\partial f}{\partial u} = 2u \dfrac{\partial f}{\partial x} + \dfrac{1}{v} \dfrac{\partial f}{\partial y}, \dfrac{\partial f}{\partial v} = 2v \dfrac{\partial f}{\partial x} - \dfrac{u}{v^2} \dfrac{\partial f}{\partial y}$

$\dfrac{\partial f}{\partial x} = \dfrac{1}{2} \dfrac{1}{u^2 + v^2} \left(u \dfrac{\partial f}{\partial u} + v \dfrac{\partial f}{\partial v} \right)$

$\dfrac{\partial f}{\partial y} = \dfrac{v^2}{u^2 + v^2} \left(v \dfrac{\partial f}{\partial u} - u \dfrac{\partial f}{\partial v} \right)$

Chapter 5

EXERCISES ON 5.2

1. If f is a real-valued function of n variables that is defined in the neighbourhood of a point $(x_1, \ldots, x_n)$ and such that all first partial derivatives exist, then f is differentiable at $(x_1, \ldots, x_n)$ if there exist functions $\varepsilon_i(\Delta x_1, \ldots, \Delta x_n)$, $i = 1, \ldots n$, such that

$$\Delta f = f(x_1 + \Delta x_1, \ldots, x_n + \Delta x_n) = \sum_{i=1}^{n} f_{x_i}(x_1, \ldots x_n)\Delta x_i + \sum_{i=1}^{n} \varepsilon_i(\Delta x_1, \ldots \Delta x_n)\, \Delta x_i$$

where

$$\lim_{(\Delta x_1, \ldots, \Delta x_n) \to (0, \ldots, 0)} \varepsilon_i(\Delta x_1, \ldots, \Delta x_n) = 0 \qquad \forall i$$

3. Not differentiable (*see* Exercise 3.3.1(i))

EXERCISES ON 5.3

2. (i) $f_x(0, 0) = \lim\limits_{(x, y) \to (0, 0)} \left[\dfrac{x}{\sqrt{x^2 + y^2}} \right]$
   (ii) Continuity does not imply differentiability

3. Not differentiable

EXERCISES ON 5.5

1. (i) $18x + 16y - z = 25$ (ii) $x - z = 0$
2. (i) $18x + 16y - 25$ (ii) x

FURTHER EXERCISES

1. (i) Differentiable   (ii) Differentiable for $x \neq y$
    (iii) Differentiable except at the origin   (iv) Differentiable
    (v) Not differentiable at the origin

4. (i) $12x - 3y - z = -16;\ 12\mathbf{i} - 3\mathbf{j} - \mathbf{k}$
    (ii) $2x - y + z = 3;\ 2\mathbf{i} - \mathbf{j} + \mathbf{k}$
    (iii) $3x - y + z = 0;\ 3\mathbf{i} - \mathbf{j} + \mathbf{k}$
    (iv) $z = 0;\ \mathbf{k}$
    (v) $z - x - y = 1;\ -\mathbf{i} - \mathbf{j} + \mathbf{k}$
    (vi) $3x + 2y + 6z = 17;\ 3\mathbf{i} + 2\mathbf{j} + 6\mathbf{k}$

5. (i) All points on the x- and y-axes   (ii) $(1, 0, -1)$ only

6. $\left(2, \dfrac{1}{2}, \dfrac{13}{4}\right)$

8. (i) (a) 1   (b) $2x - 2y - 1$
    (ii) (a) $4x - 3y + 2$ (b) $4x - 3y + 2$
    (iii) (a) 0   (b) $12x + 40y - 56$
    (iv) (a) y (b) $2x + 1$

9. (i) (a) 0   (b) $y + z$ (c) $x - y + 1$
    (ii) (a) 0   (b) $2y - 1$ (c) $2x + 2y - 2$

Chapter 6

EXERCISE ON 6.1

(i) $1 + xy + \dfrac{x^2 y^2}{2!} + \dfrac{x^3 y^3}{3!} + \ldots$

(ii) $y - \dfrac{x^2 y^3}{2!} + \dfrac{x^4 y^5}{4!} \ldots$

(iii) $x + y - \dfrac{x^3}{3!} - \dfrac{x^2 y}{2} - \dfrac{xy^2}{2} - \dfrac{y^3}{3!} + \ldots$

EXERCISES ON 6.2

1. (i) $1 + x + y + \dfrac{x^2}{2} + xy + \dfrac{y^2}{2} + \dfrac{x^3}{6} + \dfrac{x^2 y}{2} + \dfrac{xy^2}{2} + \dfrac{y^3}{6}$

    (ii) xy (iii) $1 - \dfrac{y^2}{2}$

2. (i) 0   (ii) $1 + xy + \dfrac{x^2 y^2}{2}$

    (iii) $x^2 + y^2$

EXERCISES ON 6.4

2. (i) Local minimum of -8 at $(2, 2)$. Saddle point at $(0, 0)$.

(ii) Local minimum at $\left(-\frac{1}{2}, 1\right)$, maximum at $\left(\frac{1}{2}, -1\right)$ and saddle points at $\left(\frac{1}{2}, 1\right), \left(-\frac{1}{2}, -1\right)$.

(iii) Maximum at $(-1, 0)$.

(iv) Maximum at $(0, 0)$.

3. Local minimum for $-2 < k < 2$, saddle point for $|k| > 2$, and test inconclusive if $k = \pm 2$.

EXERCISES ON 6.5

1. $\dfrac{|ax_0 + by_0 + c|}{\sqrt{a^2 + b^2}}$

2. (i) $\pm(\sqrt{2}, -1)$ both give local minima of $-\sqrt{2}$ each, $\pm(\sqrt{2}, 1)$ give maxima of $\sqrt{2}$ each.

   (ii) $\pm(1, 1)$ both local minima of 2.

   (iii) $\left(\dfrac{25}{3}, \dfrac{50}{3}, 25\right)$ gives a maximum.

   (iv) $\left(\dfrac{2}{3}, \dfrac{4}{3}, -\dfrac{4}{3}\right)$ gives a maximum of $\dfrac{4}{3}$.

FURTHER EXERCISES

1. (i) $x + xy + \dfrac{xy^2}{2} + \dfrac{xy^3}{3!} + \dots$ (ii) $x - \dfrac{xy^2}{2} + \dots$

   (iii) $x + xy + \dfrac{xy^2}{2} - \dfrac{x^3}{6} + \dots$ (iv) $1 + x^2 + y^2 + \dots$

   (v) $\ln 2 + \dfrac{x}{2} + \dfrac{y}{2} - \dfrac{x^2}{8} - \dfrac{xy}{4} - \dfrac{y^2}{8} + \dfrac{x^3}{24} + \dfrac{x^2y}{8} + \dfrac{xy^2}{8} + \dfrac{y^3}{24} + \dots$

   (vi) $1 - x - y + x^2 + 2xy + y^2 + x^3 + 3x^2y + 3xy^2 + y^3 + \dots$

   (vii) $x - x^2y + \dots$

3. (i) No critical points

   (ii) Local minimum at $(-2, -2)$

   (iii) Local minimum at $(2, 6)$

   (iv) Local minimum at $\left(-\dfrac{1}{3}, -\dfrac{1}{3}\right)$

   (v) Saddle point at $\left(-\dfrac{3}{14}, -\dfrac{1}{14}\right)$

   (vi) Minimum at $(0, 0)$ and saddle points at $(-2, 1)$ and $(2, 1)$

   (vii) Minimum at $(1, 2)$

   (viii) Maximum at $\left(\dfrac{10}{3}, \dfrac{8}{3}\right)$

4. (i) Saddle points at $\pm(1, -1)$

   (ii) Local minimum at $\left(\dfrac{-ac}{a^2 + b^2 + 1}, \dfrac{-bc}{a^2 + b^2 + 1}\right)$

   (iii) Saddle point at $(0, 0)$, minimum at $(1, 1)$

   (iv) Minimum at $(1, 2)$

   (v) Saddle points at $(n\pi, 0)$, n = integer

(vi) Minimum at $\left(1, \dfrac{3}{2}\right)$, saddle points at $(0, 1)$ and $(2, 3)$

5. (i) $\sqrt{2}\, a \times \sqrt{2}\, b$ (ii) $\dfrac{4a^2}{\sqrt{a^2 + b^2}} \times \dfrac{4b^2}{\sqrt{a^2 + b^2}}$

7. (i) Maximum of 1 at critical points $(\pm 1, 0)$, $(0, \pm 1)$

(ii) Maximum of 5 at $\left(\dfrac{3}{5}, \dfrac{4}{5}\right)$, minimum of -5 at $\left(-\dfrac{3}{5}, -\dfrac{4}{5}\right)$

(iii) Maximum of $\sqrt{2}$ at $\left(\dfrac{1}{\sqrt{2}}, 0\right)$, minimum of $-\sqrt{2}$ at $\left(-\dfrac{1}{\sqrt{2}}, 0\right)$

(iv) Maximum of 8 at $(4, 4)$

(v) Maximum of $19\sqrt{2}$ at $\left(\sqrt{2}, \dfrac{3}{\sqrt{2}}, \dfrac{5}{\sqrt{2}}\right)$

(vi) Maximum $\dfrac{1}{3\sqrt{3}}$ at $\left(\dfrac{1}{\sqrt{3}}, \dfrac{1}{\sqrt{3}}, \dfrac{1}{\sqrt{3}}\right)$, $\left(\dfrac{1}{\sqrt{3}}, -\dfrac{1}{\sqrt{3}}, -\dfrac{1}{\sqrt{3}}\right)$,

$\left(-\dfrac{1}{\sqrt{3}}, \dfrac{1}{\sqrt{3}}, -\dfrac{1}{\sqrt{3}}\right)$, $\left(-\dfrac{1}{\sqrt{3}}, -\dfrac{1}{\sqrt{3}}, \dfrac{1}{\sqrt{3}}\right)$

Minimum $-\dfrac{1}{3\sqrt{3}}$ at $\left(\dfrac{1}{\sqrt{3}}, \dfrac{1}{\sqrt{3}}, -\dfrac{1}{\sqrt{3}}\right)$, $\left(\dfrac{1}{\sqrt{3}}, -\dfrac{1}{\sqrt{3}}, \dfrac{1}{\sqrt{3}}\right)$,

$\left(-\dfrac{1}{\sqrt{3}}, \dfrac{1}{\sqrt{3}}, \dfrac{1}{\sqrt{3}}\right)$, $\left(-\dfrac{1}{\sqrt{3}}, -\dfrac{1}{\sqrt{3}}, -\dfrac{1}{\sqrt{3}}\right)$

(vii) Maximum at $\left(\dfrac{15}{2}, \dfrac{15}{2}, 15\right)$

8. $\sqrt{n}$

Chapter 7

EXERCISE ON 7.1

7/3

EXERCISE ON 7.2

7

EXERCISES ON 7.3

1. (i) $\dfrac{\pi}{5}, \dfrac{\pi}{2}$ (ii) $\dfrac{\pi}{te}, \pi$ 2. $0, 14$

EXERCISES ON 7.4

1. (i) $\dfrac{1}{24}$ (ii) $\dfrac{2}{15}$ 2. $\dfrac{1}{2}$

3. (i) $\dfrac{16}{3}$ (ii) $\dfrac{1}{3} + \dfrac{\pi}{16}$

EXERCISES ON 7.5

1. (i) $\displaystyle\int_0^{\sqrt{2}}\int_{y^2}^{2} f(x,y)\,dx\,dy$ (ii) $\displaystyle\int_{-1}^{1}\int_0^{\sqrt{1-x^2}} f(x,y)\,dy\,dx$

(iii) $\displaystyle\int_{-1}^{0}\int_{-x}^{1} f(x,y)\,dy\,dx + \int_0^{1}\int_{x}^{1} f(x,y)\,dy\,dx$

(iv) $\displaystyle\int_{-1}^{e^2}\int_{\ln y}^{2} f(x,y)\,dx\,dy$

(v) $\displaystyle\int_1^{2}\int_2^{4} f(x,y)\,dy\,dx + \int_2^{4}\int_{x}^{4} f(x,y)\,dy\,dx$

2. (i) 1 (ii) $1 - \ln 2$ (iii) 27

3. (i) $1 - \cos 1$ (ii) $\dfrac{1}{\pi}$

EXERCISES ON 7.6

1. 4 **2.** $4\pi a^3/3$ **3.** 1/6

4. (i) $\dfrac{\pi}{4}\left(1 - \cos 1\right)$ (ii) $\dfrac{\pi}{2}\left(1 - \dfrac{1}{\sqrt{5}}\right)$

EXERCISE ON 7.7

$\dfrac{1}{12}(5\sqrt{5} - 1)$

EXERCISES ON 7.8

1. 180 **2.** 1/8 **3.** $243\pi/2$ **4.** $\pi k a^4$

EXERCISES ON 7.9

1. $\dfrac{1}{2}(e^2 - e)\ln 2$

2. (i) r (ii) $\rho^2 \sin \phi$

FURTHER EXERCISES

1. (i) (a) 1 (b) $-\dfrac{3}{4}$

(ii) (a) 5 (b) $-\dfrac{3}{2}$

(iii) (a) 4   (b) $\dfrac{3}{2}$

(iv) (a) $\dfrac{3}{2}e^2 - \dfrac{9}{2}$ (b) $\dfrac{1}{2}e^{-2} - 2e + \dfrac{9}{2}$

(v) (a) $\dfrac{17}{6}$ (b) $\dfrac{47}{4}$

2. (i) 1   (ii) $-\dfrac{7}{8}$ (iii) $\dfrac{8}{3}$

(iv) $\dfrac{1}{3}(e - e^2)$

3. $\dfrac{\pi}{2}, \pi$

4. (i) $\dfrac{18}{105}$ (ii) $\dfrac{76}{15}$ (iii) $\dfrac{4}{7}$

(iv) $\dfrac{11}{6}$ (v) $-\dfrac{68}{3}$ (vi) $\ln 3$

(vii) $\dfrac{1}{2}$ (viii) $\dfrac{11}{24}$

5. (i) 1   (ii) $\dfrac{1}{7}$ (iii) $e - 2$ (iv) $4\ln 2 - \dfrac{3}{2}$

6. (i) 2   (ii) 1   (iii) $\dfrac{1}{4}(e^8 - 1)$ (iv) $\ln 2$

7. (i) $\dfrac{292}{3}$ (ii) $-\dfrac{68}{3}$ (iii) $\dfrac{4}{3}$ (iv) $\dfrac{625}{12}$

8. (i) 1   (ii) π^2

9. (i) $\dfrac{\pi}{4}$ (ii) πa^2 (iii) $(1 - \ln 2)\pi$ (iv) $\dfrac{\pi}{2} + 1$

10. 2π

11. $\dfrac{3}{8}\pi + 1$

12. $\dfrac{\sqrt{\pi}}{2}$

13. (i) 1   (ii) 1   (iii) 0   (iv) $8\ln 8$

14. (i) $\dfrac{2}{3}$ (ii) $\dfrac{1}{6}$

15. (i) $\dfrac{a^2}{3}$ (ii) $\dfrac{1}{6}$

16. (i) $\displaystyle\int_0^1 \int_{y^{1/2}}^{y^{1/4}} f(x, y)\, dx\, dy$ (ii) $\displaystyle\int_{-1}^0 \int_{-x}^1 f(x, y)\, dy\, dx + \int_0^1 \int_x^1 f(x, y)\, dy\, dx$

17. (i) 4   (ii) $\dfrac{1}{3}$

18. (i) 4π (ii) $\dfrac{1}{3}$

19. $\dfrac{8\pi\sqrt{2}}{3}$

20. (i) $\dfrac{\pi}{3}$　　(ii) $\dfrac{\pi}{2}$

22. 8π

23. (i) $ad - bc$　　(ii) $\cos(u - v)$　　(iii) uvw^2　　(iv) $\dfrac{1}{4}\sin 2r \sin 2s \sin 2t$

26. $-\dfrac{2}{3}\ln 2$　　**27.** $\dfrac{2}{3}$

Chapter 8

EXERCISES ON 8.3

1. (i) 1　　　　　(ii) 10　　　　(iii) −1　　　　(iv) −1
　　(v) 3　　　　　(vi) 40　　　　(vii) 27

2. (i) $\mathrm{d}\phi = \dfrac{\partial\phi}{\partial x_i}\mathrm{d}x_i$　　(ii) $x_i x_i$　　(iii) $a_{ji}b_i$

EXERCISES ON 8.4

2. $(2, 1, 1)$

EXERCISES ON 8.5

1. (i) $x = t, y = t^2, z = t^3$　　　　(ii) $x = e^t, y = -2t, z = \cos t$

2.

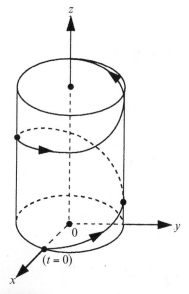

EXERCISE ON 8.6

　　(i) $t(t^2 - 1) + 6t^2 \cos t - te^t$
　　(ii) $2t\mathbf{i} - 3(\sin t)\mathbf{j} - \mathbf{k}$
　　(iii) $(2t^2 + t - 2)\mathbf{i} + (2t^2 + 6 \cos t)\mathbf{j} + (e^t - 2t)\mathbf{k}$
　　(iv) $-(2t^3 + 3e^t \cos t)\mathbf{i} + (t^2 + e^t(t^2 - 1))\mathbf{j} + (3t \cos t + 2t^2(t^2 - 1))\mathbf{k}$

(v) $3t^2 - 1 + 6t(2 \cos t - t \sin t) + e^t(t-1)$

(vi) $(4t + 1)\mathbf{i} + 2(2t - 3 \sin t)\mathbf{j} + (e^t - 2)\mathbf{k}$

(vii) $-(6t^2 + 3e^t(\cos t - \sin t))\mathbf{i} + (2t + e^t(t^2 + 2t - 1))\mathbf{j}$
$+ (3 \cos t - 3t \sin t + 8t^3 - 4t)\mathbf{k}$

(viii) $6 \cos 1 - 12 \sin 1 + e + \dfrac{39}{4}$

(ix) $-\dfrac{8}{3}\mathbf{i} + \left(12 \sin 1 + \dfrac{4}{3}\right)\mathbf{j} + (e - e^{-1})\mathbf{k}$

(x) $\dfrac{t + 8t^3 + e^{2t}}{\sqrt{t^2 + 4t^4 + e^{2t}}}$

EXERCISE ON 8.7

(i) $\mathbf{T} = \dfrac{-(\sin t)\mathbf{i} + (\cos t)\mathbf{j} + \mathbf{k}}{\sqrt{2}}$, $\mathbf{N} = -(\cos t)\mathbf{i} - (\sin t)\mathbf{j}$

(ii) $\mathbf{T} = \dfrac{t^2\mathbf{i} + \sqrt{2}t\mathbf{j} + \mathbf{k}}{t^2 + 1}$, $\mathbf{N} = \dfrac{\sqrt{2}t\mathbf{i} + (1 - t^2)\mathbf{j} - \sqrt{2}t\mathbf{k}}{t^2 + 1}$

EXERCISE ON 8.8

(i) $\dfrac{9}{2}(82^{3/2} - 1)$ (ii) $\sqrt{6}\,(e^4 - e)$

EXERCISES ON 8.9

$\mathbf{T} = \dfrac{-a(\sin t)\mathbf{i} + a(\cos t)\mathbf{j} + b\mathbf{k}}{\sqrt{a^2 + b^2}}$

$\mathbf{N} = -(\cos t)\mathbf{i} - (\sin t)\mathbf{j}$

$\mathbf{B} = \dfrac{b(\sin t)\mathbf{i} - b(\cos t)\mathbf{j} + a\mathbf{k}}{\sqrt{a^2 + b^2}}$

FURTHER EXERCISES

1. (i) Rotation about the z-axis through an angle of $60°$ from the x-axis towards the y-axis.
    (ii) $30°$ rotation about x-axis from the y- to the z-axis.
    (iii) Rotation of $45°$ about the y-axis from z- to x-axis.

3. (i) $a_i x_i x_i$ (ii) $\dfrac{\mathrm{d}y_k}{\mathrm{d}t} = \dfrac{\partial y_k}{\partial x_i}\dfrac{\mathrm{d}x_i}{\mathrm{d}t}$

4. m is a scalar.

5. (i) (a) $-(\sin t)\mathbf{i} + (\cos t)\mathbf{j} + \mathbf{k}$ (b) $-(\cos t)\mathbf{i} - (\sin t)\mathbf{j}$
      (c) $\sqrt{2}$ (d) 1
    (ii) (a) $e^t\mathbf{i} - 4e^{-2t}\mathbf{j} + (\sin t)\mathbf{k}$ (b) $e^t\mathbf{i} + 8e^{-2t}\mathbf{j} + (\cos t)\mathbf{k}$
      (c) $\sqrt{e^{2t} + 16e^{-4t} + \sin^2 t}$ (d) $\sqrt{e^{2t} + 64e^{-4t} + \cos^2 t}$
    (iii) (a) $\mathbf{i} - 4t\mathbf{j} + 3\mathbf{k}$ (b) $-4\mathbf{j}$
      (c) $\sqrt{10 + 16t^2}$ (d) 4
    (iv) (a) $-(\sin t)\mathbf{i} - (\cos t)\mathbf{j} - 2(\sin 2t)\mathbf{k}$ (b) $-(\cos t)\mathbf{i} + (\sin t)\mathbf{j} - 4(\cos 2t)\mathbf{k}$
      (c) $\sqrt{1 + 4 \sin^2 2t}$ (d) $\sqrt{1 + 16 \cos^2 2t}$

6. (i) $3t^2 - 10t$ (ii) $-2\mathbf{i} + 4\mathbf{j}$

(iii) $7t^2\mathbf{i} - (6t^2 + t)\mathbf{j} + (2t^2 - 2t^3)\mathbf{k}$

(iv) $(3t^2 - 6)\mathbf{i} - (9t^2 + 2t)\mathbf{j} - (4t^3 + 2)\mathbf{k}$

(v) $\dfrac{t^4}{4} - \dfrac{5t^3}{3} + C$

(vi) $\left(\dfrac{t^4}{4} - 3t^2\right)\mathbf{i} - \left(\dfrac{3}{4}t^4 + \dfrac{t^3}{3}\right)\mathbf{j} - \left(\dfrac{t^5}{5} + t^2\right)\mathbf{k} + \mathbf{C}$

(vii) $-\dfrac{3}{10}$ (viii) $-8\mathbf{i} - \dfrac{44}{3}\mathbf{j} - \dfrac{52}{5}\mathbf{k}$

(ix) $\dfrac{4t^7 + 30t^5 + 25t^4 - 22t^3 + 40t}{t^8 + 10t^6 + 10t^5 - 11t^4 + 40t^2}$

(x) $\dfrac{4 - t^4}{(t^4 + t^2 + 4)^{3/2}}\mathbf{i} + \dfrac{t^3 + 8t}{(t^4 + t^2 + 4)^{3/2}}\mathbf{j} + \dfrac{3(t^4 - 4)}{(t^4 + t^2 + 4)^{3/2}}\mathbf{k}$

8. $\mathbf{v} \cdot \mathbf{v}' \times \mathbf{v}'''$

10. (i) $-(\sin t)e^{\cos t}\mathbf{i} + (\cos t)e^{\sin t}\mathbf{j} + \mathbf{k};$

$e^{\cos t}(\sin^2 t - \cos t)\mathbf{i} + e^{\sin t}(\cos^2 t - \sin t)\mathbf{j}$

(ii) $-e^t(\cos t + \sin t);\ -2e^t\cos t$

(iii) $\mathbf{b} \times \mathbf{c} + \mathbf{a} \times \mathbf{d} + 2\mathbf{b} \times \mathbf{d}t;\ 2\mathbf{b} \times \mathbf{d}$

(iv) $2\mathbf{a}\cdot\mathbf{b} + 2(\mathbf{c}\cdot\mathbf{a} + \mathbf{b}^2)t + 3\mathbf{b}\cdot\mathbf{c}t^2;\ 2(\mathbf{a}\cdot\mathbf{c} + \mathbf{b}^2) + 6(\mathbf{b}\cdot\mathbf{c})t$

12. $(-1, 1, 2)$

16. $\dfrac{\partial \mathbf{F}}{\partial x} = 2xy\mathbf{i} + y^2z\mathbf{j} + (\cos(xz) - xz\sin(xz))\mathbf{k}$

$\dfrac{\partial \mathbf{F}}{\partial y} = x^2\mathbf{i} + 2xyz\mathbf{j}$

$\dfrac{\partial \mathbf{F}}{\partial z} = -2\mathbf{i} + xy^2\mathbf{j} - x^2\sin(xz)\mathbf{k}$

$\dfrac{\partial^2 \mathbf{F}}{\partial x^2} = 2y\mathbf{i} - (2z\sin(xz) + xz^2\cos(xz))\mathbf{k}$

$\dfrac{\partial^2 \mathbf{F}}{\partial y^2} = 2xz\mathbf{j}$ $\dfrac{\partial^2 \mathbf{F}}{\partial z^2} = -x^3\cos(xz)\mathbf{k}$

$\dfrac{\partial^2 \mathbf{F}}{\partial x\partial y} = 2x\mathbf{i} + 2yz\mathbf{j}$ $\dfrac{\partial^2 \mathbf{F}}{\partial y\partial z} = 2xz\mathbf{j}$

$\dfrac{\partial^2 \mathbf{F}}{\partial x\partial z} = -(2x\sin(xz) + x^2z\cos(xz))\mathbf{k}$

17. (i) $\mathbf{T} = \dfrac{3}{5}(\sin t)\mathbf{i} + \dfrac{3}{5}(\cos t)\mathbf{j} + \dfrac{4}{5}\mathbf{k}$

(ii) $\mathbf{N} = -(\cos t)\mathbf{i} - (\sin t)\mathbf{j},\ \kappa = \dfrac{3}{25},\ \rho = \dfrac{25}{3}$

(iii) $\mathbf{B} = \dfrac{4}{5}(\sin t)\mathbf{i} - \dfrac{4}{5}(\cos t)\mathbf{j} + \dfrac{3}{5}\mathbf{k},\ \tau = \dfrac{4}{25}$

20. (i) (a) $\mathbf{T} = \dfrac{1 + 2t\mathbf{j} + 2t^2\mathbf{k}}{1 + 2t^2}$

$\kappa = \dfrac{2}{(1 + 2t^2)^2}$

(b) $\mathbf{N} = \dfrac{-2t\mathbf{i} + (1 - 2t^2)\mathbf{j} + 2t\mathbf{k}}{1 + 2t^2}$

$$\mathbf{B} = \frac{2t^2\mathbf{i} - 2t\mathbf{j} + \mathbf{k}}{1 + 2t^2}$$

$$\tau = \frac{2}{(1 + 2t^2)^2}$$

(ii) (a) $\mathbf{T} = \dfrac{e^t\mathbf{i} - e^{-t}\mathbf{j} + \sqrt{2}\,\mathbf{k}}{e^t + e^{-t}}$

$$\kappa = \frac{\sqrt{2}}{(e^t + e^{-t})^2}$$

(b) $\mathbf{N} = \dfrac{\sqrt{2}}{e^t + e^{-t}}\mathbf{i} + \dfrac{\sqrt{2}}{e^t + e^{-t}}\mathbf{j} - \dfrac{e^t - e^{-t}}{e^t + e^{-t}}\mathbf{k}$

$$\mathbf{B} = \frac{-e^{-t}}{e^t + e^{-t}}\mathbf{i} + \frac{e^t}{e^t + e^{-t}}\mathbf{j} + \frac{\sqrt{2}}{e^t + e^{-t}}\mathbf{k}$$

$$\tau = -\frac{\sqrt{2}}{(e^t + e^{-t})^2}$$

(iii) (a) $\mathbf{T} = \dfrac{1}{\sqrt{2}}\left(\dfrac{1 - t^2}{1 + t^2}\right)\mathbf{i} + \dfrac{\sqrt{2}\,t}{1 + t^2}\mathbf{j} + \dfrac{1}{\sqrt{2}}\mathbf{k}$

$$\kappa = \frac{1}{(1 + t^2)^2}$$

(b) $\mathbf{N} = -\dfrac{2t}{1 + t^2}\mathbf{i} + \left(\dfrac{1 - t^2}{1 + t^2}\right)\mathbf{j}$

$$\mathbf{B} = \frac{(t^2 - 1)\mathbf{i} - 2t\mathbf{j} + (1 + t^2)\,\mathbf{k}}{\sqrt{2}(1 + t^2)}$$

$$\tau = \frac{1}{(1 + t^2)^2}$$

Chapter 9

EXERCISES ON 9.1

1. Increases (decreases) most rapidly in direction $\mathbf{j}$ ($-\mathbf{j}$). Zero rate of change in directions $\pm\mathbf{i}$.

2. (i) $\pm\left(\dfrac{1}{\sqrt{11}}\mathbf{i} - \dfrac{3}{\sqrt{11}}\mathbf{j} - \dfrac{1}{\sqrt{11}}\mathbf{k}\right); \pm\sqrt{11}$

(ii) $\mp\left(\dfrac{4}{3\sqrt{2}}\mathbf{i} - \dfrac{1}{3\sqrt{2}}\mathbf{j} - \dfrac{1}{3\sqrt{2}}\mathbf{k}\right); \pm 3\sqrt{2}$

EXERCISES ON 9.2

1. (i) $2x + 2y + 4\sqrt{2}z = 5$

$$x = \frac{1}{4} + \frac{1}{2}t, \quad y = \frac{1}{4} + \frac{1}{2}t, \quad z = \frac{1}{\sqrt{2}} + \sqrt{2}t$$

(ii) $y = \dfrac{\pi}{2}$

$$x = 0, \quad y = \frac{\pi}{2} - \frac{1}{2}t, \quad z = \frac{\pi}{3}$$

2. (i) $2y - z = 1$ (ii) $x + y - z = -1$

EXERCISE ON 9.5

(i) Parallel vectors at each point at 45° to the *x*-axis, with length $K\sqrt{2}$.

(ii) Vectors at each point except the origin radiating out from the origin.

EXERCISES ON 9.6

1. (i) $2x\mathbf{i} + 6y\mathbf{j}$ (ii) $2(x + y + z)(\mathbf{i} + \mathbf{j} + \mathbf{k})$

(iii) $-y \sin(xy)\mathbf{i} - x \sin(xy)\mathbf{j}$

(iv) $ze^x(\cos y)\mathbf{i} - ze^x(\sin y)\mathbf{j} + e^x(\cos y)\mathbf{k}$

(v) $\dfrac{(1 - y^2 + xz)}{(1 - y^2 + x^2)^{3/2}}\mathbf{i} - \dfrac{y(x - z)}{(1 - y^2 + x^2)^{3/2}}\mathbf{j} - \dfrac{1}{\sqrt{1 - y^2 + x^2}}\mathbf{k}$

2. $\dfrac{4xy}{((x - 1)^2 + y^2)\,((x + 1)^2 + y^2)}\mathbf{i} + \dfrac{2(x^2 - y^2 - 1)}{((x - 1)^2 + y^2)((x + 1)^2 + y^2)}\mathbf{j}$

3. $\phi = xyz + \text{constant}$

EXERCISES ON 9.7

2. (i) (a) 4   (b) 0   (c) 0

(ii) (a) $e^x + e^y + e^z$ (b) 0   (c) F

(iii) (a) $\dfrac{\partial F_1}{\partial x} + \dfrac{\partial F_2}{\partial y} + \dfrac{\partial F_3}{\partial z}$ (b) 0

(c) $\dfrac{\partial^2 F_1}{\partial x^2}\mathbf{i} + \dfrac{\partial^2 F_2}{\partial y^2}\mathbf{j} + \dfrac{\partial^2 F_3}{\partial z^2}\mathbf{k}$

(iv) (a) $-y \sin(xy) + z \cos(yz)$

(b) $y \cos(yz)\mathbf{i} - x \sin(xy)\mathbf{k}$

(c) $-y^2 \cos(xy)\mathbf{i} - (\sin(xy) + xy \cos(xy) + z^2 \sin(yz))\mathbf{j}$
$+ (\cos(yz) - y \sin(yz))\mathbf{k}$

All other combinations of div, curl, grad are either identically zero or are not defined on a vector.

FURTHER EXERCISES

1. (i) $2\sqrt{2}$ (ii) $\dfrac{e^2}{\sqrt{13}}(8 \cos 3 - 5 \sin 3)$

(iii) $\dfrac{1}{\sqrt{3}}$ (iv) $-\dfrac{2}{\sqrt{42}}$

2. (i) $2x\mathbf{i} + 4y\mathbf{j} + 6z\mathbf{k}$

(ii) $yz(\sin(x + y + z) + x \cos(x + y + z))\mathbf{i}$
$+ xz(\sin(x + y + z) + y \cos(x + y + z))\mathbf{j}$
$+ xy(\sin(x + y + z) + z \cos(x + y + z))\mathbf{k}$

(iii) $e^{xyz}\left(yz \ln(x + y + z) + \dfrac{1}{x + y + z}\right)\mathbf{i}$
$+ e^{xyz}\left(xz \ln(x + y + z) + \dfrac{1}{x + y + z}\right)\mathbf{j}$
$+ e^{xyz}\left(xy \ln(x + y + z) + \dfrac{1}{x + y + z}\right)\mathbf{k}$

(iv) $\dfrac{\mathbf{r}}{|\mathbf{r}|}$

(v) $(3y^2 + 4xz + 4yz)\mathbf{i} + (6xy + 4xz)\mathbf{j} + (2x^2 + 4xy)\mathbf{k}$

(vi) $-y\sin(xy)\mathbf{i} + (z\cos(yz) - x\sin(xy))\mathbf{j} + y\cos(yz)\mathbf{k}$

3. (i) (a) $\dfrac{\mathbf{i} + 3\mathbf{j}}{\sqrt{10}}$ (b) $-\dfrac{(\mathbf{i} + 3\mathbf{j})}{\sqrt{10}}$

   (ii) $\pm\left[\dfrac{(1 - 2\sin 4)\mathbf{i} + (2 - 2\sin 4)\mathbf{j} - 2\,(\sin 4)\mathbf{k}}{\sqrt{5 - 12\sin 4 + 12\sin^2 4}}\right]$

   (iii) $\pm\dfrac{1}{\sqrt{3}}(\mathbf{i} + \mathbf{j} + \mathbf{k})$

4. 0.0008

5. (i) $x + z = 2$
   $$x = 1 + 2t, \qquad y = 0, \qquad z = 1 + 2t$$

   (ii) $x - 2y + z = 0$
   $$x = 2 + 2t, \qquad y = 2 - 4t, \qquad z = 0$$

   (iii) $6x + 3y + 2z = 18$
   $$x = 1 + 2t, \qquad y = 2 + t, \qquad z = 3 + \frac{2}{3}t$$

   (iv) $2x + y + z = 6$
   $$x = 1 + 4t,\ y = 2 + 2t,\ z = 2 + 2t$$

   (v) $4x + 4y - z = 8$
   $$x = 2 - 4t, \qquad y = 1 - 4t, \qquad z = 4 + t$$

   (vi) $3y - z + 1 = 0$
   $$x = \frac{\pi}{6}, \qquad y = 3t, \qquad z = 1 - t$$

6. (i) $3r\underline{\mathbf{r}}$ (ii) $\dfrac{f'(r)}{r}\mathbf{r}$

7. $\phi(r) = \dfrac{1}{3}\left(\dfrac{1}{2^3} - \dfrac{1}{r^3}\right)$

8. $\phi = x^2 yz^3 + 20$

11. (i) $0,\ 2\omega\mathbf{k}$ (ii) $2,\mathbf{0}$

    (iii) $yz + 1,\qquad -x\mathbf{i} + xy\mathbf{j} + (1 - x)z\mathbf{k}$

    (iv) $4,\qquad -x^2\,z\mathbf{i} + 8xyz\mathbf{j} + (2y^3 - 3xz^2)\mathbf{k}$

    (v) $6xy + 3y^2,\qquad \mathbf{0}$

    (vi) $x + y,\qquad z\mathbf{j} + x\mathbf{k}$

    (vii) $f'(r) + 2\dfrac{f'(r)}{r},\ \mathbf{0}$

    (viii) $y - y\sin(yz) - 2,\qquad (-z\sin(yz) - 2z)\mathbf{i} - x\mathbf{k}$

    (ix) $4xz - 2xyz + 6yz,\qquad (xy^2 + 3z^2)\mathbf{i} - 2x^2\mathbf{j} - y^2z\mathbf{k}$

12. (i) 1   (ii) $2\mathbf{i} + 2\mathbf{j}$

    (iii) -5 (iv) 2

    (v) $\mathbf{i} - \mathbf{j} - \mathbf{k}$ (vi) $-2\mathbf{i} - 2\mathbf{j}$

    (vii) $-2\mathbf{i} - 2\mathbf{j}$

13. (i) 0   (ii) 0

    (iii) $6y + 2y^3z + 6x^2yz$ (iv) $-e^{-x}\cos y \sin z$

    (v) 0   (vi) $\dfrac{1}{r^2}$

    (vii) $n(n + 1)r^{n-2}$

14. $f(r) = A + \dfrac{B}{r}$, A, B arbitrary constants

Chapter 10

EXERCISES ON 10.2

1. 2

2. (i) $\dfrac{\pi^3}{24}$ (ii) 1

EXERCISES ON 10.3

1. (i) 0   (ii) −2
The zero result (i) reflects the fact that the line integral given is the integral of a total differential, $d[(x − y)^2/2]$ around a closed contour. This is also revealed by the fact that $\partial Q/\partial y = \partial P/\partial x$ in this case.

EXERCISES ON 10.4

2. $\dfrac{4\pi}{3}$ **3.** $\dfrac{3\pi}{2}$

EXERCISES ON 10.6

1. 4π

FURTHER EXERCISES

1. (i) $\dfrac{21}{2}$ (ii) $\dfrac{65}{6}$ (iii) $\dfrac{38}{3}$

2. (i) $-\left(\dfrac{254}{3} + \ln 27\right)$ (ii) 81   (iii) 1

   (iv) 2π (v) $\dfrac{129}{4}$ (vi) 3

3. (i) 6   (ii) $\dfrac{8\pi^3}{3}$ (iii) $\dfrac{579}{35}$

4. -36π

5. (i) $\phi = xy^2 - e^y + C$; $1 - e$ (ii) $\phi = x^2y^2 + y^2 + C$; 80
   (iii) $\phi = xy^2z^4 + C$; 12   (iv) $\phi = x^2y - 3yz + 2xz^2 + C$; 6
   (v) $\phi = 2x^2y - x^3z^2 + C$; 2

6. $\phi = -\dfrac{r^4}{4} + C$

7. Not conservative

8. (i) 0   (ii) $\dfrac{3\pi}{4}$ (iii) 1   (iv) $\dfrac{1}{6}$

9. (i) -2π (ii) a^2b (iii) -4 (iv) 4

    (v) $-16\sqrt{2}\,\pi$ (vi) 60π (vii) $-\dfrac{7}{60}$ (viii) $\dfrac{1}{2}$

10. (i) 216 (ii) 171 (iii) $\dfrac{12}{5}\pi a^5$ (iv) 0

    (v) 4 (vi) 0 (vii) 0 (viii) $2\pi\ell a^2$

    (ix) 132

11. (i) $\dfrac{15\pi}{\sqrt{2}}$ (ii) $\dfrac{3\pi}{4}$ (iii) $\dfrac{\pi}{4}[3\sqrt{2}-\ln(\sqrt{2}+1)]$

    (iv) $\dfrac{\pi}{2\sqrt{2}}$ (v) $\dfrac{2\sqrt{14}}{3}$ (vi) $3\sqrt{2}$

12. (i) π (ii) $-\pi$

    P and Q do not have continuous derivatives throughout any region containing $(0,0)$.

13. (i) $\dfrac{5\pi a^4}{4}$ (ii) 24π (iii) 2 (iv) 180

14. (i) -16 (ii) 32π (iii) 24π (iv) $3\pi a^2$

    (v) 0 (vi) 64π

16. (i) -4 (ii) -3π (iii) 12π (iv) $\dfrac{3\pi}{2}$

    (v) $\dfrac{32}{3}$

17. (i) $-\dfrac{5}{6}$ (ii) 5π (iii) 0 (iv) π

18. (i) -16π (ii) -4π

Index